Frederick William Pavy

The physiology of the carbohydrates

Their application as food and relation to diabetes

Frederick William Pavy

The physiology of the carbohydrates
Their application as food and relation to diabetes

ISBN/EAN: 9783337201395

Printed in Europe, USA, Canada, Australia, Japan

Cover: Foto ©berggeist007 / pixelio.de

More available books at **www.hansebooks.com**

THE PHYSIOLOGY OF THE CARBOHYDRATES;

THEIR APPLICATION AS FOOD AND RELATION TO DIABETES.

BY

F. W. PAVY, M.D, LL.D., F.R.S.,

FELLOW OF THE ROYAL COLLEGE OF PHYSICIANS;
CONSULTING PHYSICIAN TO, AND FORMERLY LECTURER ON PHYSIOLOGY AND ON THE PRACTICE OF MEDICINE AT, GUY'S HOSPITAL.

(ALL RIGHTS OF TRANSLATION AND REPRODUCTION RESERVED.)

LONDON:
J. & A. CHURCHILL,
11, NEW BURLINGTON STREET.

1894.

LONDON:
HARRISON AND SONS, PRINTERS IN ORDINARY TO HER MAJESTY,
ST. MARTIN'S LANE.

PREFACE.

The carbohydrate principles constitute by far the largest portion of organic matter. They thus hold a position of great significance in relation to living nature. Taking origin in the vegetable kingdom as a result of the operation of solar influence through living matter, they primarily play a part in the scheme of life of the vegetable organism, and secondarily enter, either directly or indirectly, into the food supply of animals.

It is this latter subject which is specially dealt with in this volume, but a comprehensive view of the bearings of the carbohydrates throughout both living kingdoms requires to be taken, in order that a right understanding may be obtained of their application within the animal system. The prevailing notions have been constructed upon a fallacious foundation. I have entered minutely into the experimental evidence by which the error existing is made manifest. By the glycogenic doctrine the mind has been conducted in the wrong direction, and, as a natural consequence, the search for knowledge has hitherto resulted only in fruitless gropings in the dark. Through the recognition of the glucoside constitution of proteid matter a clue was given which has led to the discovery of what I venture to regard as the true key to the situation. With the light that has been thrown upon the question, nature can be intelligibly read, and we have simply to look around and take notice of the results that are lying clearly open to view before us. The new departure brings the operations of animal and plant life into harmony with each other. The simplest of organisms—a yeast cell for example—may, indeed, be made use of to illustrate the occurrences taking place in our own bodies. Moreover, the teachings of experience in connection with diabetes, which may be pronounced to be absolutely irreconcilable with the prevailing physiological views, not only agree with, but in the strongest manner support, the doctrine that is in this volume set forth.

A life's labour, attended with unceasing laboratory work, has been devoted to the attainment of the knowledge that has been acquired. To the authorities of Guy's Hospital I shall ever feel grateful for the assistance rendered in the cause of science by the provision of laboratory accommodation throughout the period of my association as a member of the acting staff. Since that period I have found in the research laboratories of the Royal Colleges of Physicians and Surgeons at the Examination Hall everything that could be desired for facilitating my experimental work, and thus promoting advance to the goal that has now been reached.

During the many years that my investigations have been carried on I have been assisted by a succession of zealous and able helpers, derived from past students of the Royal College of Chemistry. Notable amongst the assistants who have rendered me valuable service are Mr. Scard, Mr. Rowntree, and Mr. Siau. It is a pleasing duty to tender this acknowledgment, and it remains for me further to say that to Mr. Rowntree I am, in addition, indebted for painstaking aid in the labour involved in the issue of this work.

35, *Grosvenor Street,*
Grosvenor Square.
May, 1894.

TABLE OF CONTENTS.

PAGE

INTRODUCTORY CONSIDERATIONS. Definition of term carbohydrate. Genesis of carbohydrate matter. Chemical constitution of carbohydrates. Chemical characters and relations considered under groupings into amyloses, saccharoses, and glucoses. The amylose carbohydrates—cellulose—starch—glycogen—animal gum—dextrin. The saccharoses—maltose—saccharose or cane sugar—lactose or milk sugar. The glucoses—dextrose or grape sugar—lævulose—galactose 1

BEHAVIOUR OF SUGARS WITH PHENYL-HYDRAZINE. Osazones 17

TRANSMUTATION OF CARBOHYDRATES. Transmutation by increased hydration—effect of acids—effect of ferment action. Transmutation by decreased hydration. Diminution of hydration by artificial means. Diminution of hydration by living action—in yeast cell—in sugar cane—in liver. Transmutation by ferment and protoplasmic actions considered in relation to the operations of life. Amylolytic or diastatic ferments. Proteolytic or peptonising ferments. Synthetic power of living protoplasm. Alternate play of ferment and protoplasmic actions 18

GLUCOSIDES. Nature and examples of glucosides. The glucoside constitution of proteid matter. Discovery of the cleaving off of carbohydrate from proteid—mode of experimenting—effect of different strengths of potash—proteids experimented with—traceable effect produced by super-heated water. Preparation and properties of the primary non-reducing cleavage product by potash from proteid matter. Strength of alcohol required for precipitation. Resemblance of non-reducing product to Landwehr's animal gum. Its diffusibility. Preparation and properties of the cupric oxide reducing product (sugar) derivable, through the primary cleavage product, from proteid matter. Effects produced by 2 per cent. and by 10 per cent. sulphuric acid on the primary cleavage product. Reactions of the sugar product with Fehling's solution, phenyl hydrazine, lead oxide, &c. Its negative behaviour in relation to fermentation and optical activity. Reference to other optically inactive and unfermentable sugars. Susceptibility of the primary non-reducing product of being thrown down in combination with copper oxide. Cleavage of carbohydrate from proteid by direct action of sulphuric acid—mode of procedure—character of osazone subsequently given with phenyl-hydrazine. Suggested unrecognised modifications of sugars. Recovery of the cleavage sugar from its osazone—melting point of the osazone. Cleavage of carbohydrate from proteid by digestive ferment action. Conclusion from whole array of evidence presented. Synthetic formation by protoplasmic agency of proteids by incorporation of carbohydrate with other matter. View held with regard to participation of asparagin. Proteid matter in relation to the deposition of carbohydrate as starch, cellulose, &c., in the vegetable organism 27

DESCRIPTION OF ANALYTICAL STEPS OF PROCEDURE. Modes of extracting sugar from animal structures. The sodium sulphate method. Alcohol method preferable. Employment of sulphuric acid to obtain information regarding the nature of the sugar present. Details of the alcoholic process of extraction in the case of blood. Process for separating and estimating the amylose from the other forms of carbohydrate by the successive employment of potash, alcohol, sulphuric acid, and the ammoniated cupric test. Mode of calculating the results. Application of the process to solid tissues. Difficulty arising from glycogen and cleavage carbohydrate being yielded by the same process. Necessity of securing that the sulphuric acid used for conversion into glucose exists as free acid. Time required for conversion of the various carbohydrates into glucose. Use of the autoclave for facilitating the conversion. Use of citric acid for inversion of cane sugar. Qualitative and quantitative testing—cupric oxide reducing power the basis. Composition of Fehling's solution. Modification of Fehling's solution. Quantitative methods. Gravimetric copper process. Volumetric method—the ammoniated cupric test—the apparatus—precautions to be observed. Experimental testimony to the reliability of the test 58

INGESTED CARBOHYDRATES TRACED TO THE PORTAL BLOOD. Starch--necessity for a process of digestion—action of saliva—maltose the end product—effect of gastric juice—absorption from stomach—action of bile and pancreatic juice—modifying influence of acids and carbonated alkalis on pancreatic ferment action—action of the succus entericus or intestinal juice—its glucose-forming capacity—product of starch digestion found in portal blood—large presence of sugar with lower cupric oxide reducing power than that of glucose. Cellulose—its resistance to solvents—its susceptibility of yielding to the action of certain ferments—question of how far this occurs in the alimentary tract. Cane sugar—its position as a diffusible body—its inversion by acids—presence of an inverting ferment to a slight extent in the walls of the stomach, and to a much larger extent in those of the intestine—continued activity of the ferment after precipitation by alcohol, and drying—contrast in the case of the ruminant animal in relation to the inverting ferment—inverting effect of contact with the stomach contents—inverting action of dilute acids at moderate temperatures—product from ingested cane sugar found in the portal blood. Lactose—its position as a diffusible body—its alleged conversion into dextrose and galactose by the succus entericus—its probable partial conversion into lactic acid. Glucose—its absorption as such—possible occurrence of a certain amount of lactic acid fermentation 81

TABLE OF CONTENTS. vii

PAGE

PORTAL BLOOD IN RELATION TO INGESTED CARBOHYDRATES. Amount of sugar in blood of general circulation. Procedure for obtaining portal blood. Cupric oxide reducing power of sugar of portal blood and that of other blood. Amount and nature of sugar in portal blood—after fasting—after animal food—after starchy food—after maltose—and after cane sugar 101

THE LIVER IN RELATION TO THE SUGAR DERIVED FROM INGESTED CARBOHYDRATES. Evidence afforded through the Blood of Sugar being stopped by the Liver. Comparative states of the blood of the portal and hepatic veins. Precautions requisite in collecting blood for analysis to avoid fallacy from *post-mortem* change. Amounts of sugar encountered immediately after death in portal and hepatic blood. Sugar arresting function of liver the protection against diabetes. Evidence afforded through the Liver itself of the stoppage by it of sugar derived from Ingestion. Bernard's allegation of a sugar-forming function of the liver. His discovery of glycogen. Ingested carbohydrate a source of glycogen. Glycogen also formed from ingested animal food. Both free and combined carbohydrate in animal matter. Proteid cleavage the work of digestive ferments not of the liver. General relations of glycogen as a constituent of the liver. Not easily extracted. Amount in the liver. Its diminution from *post-mortem* transformation. Range of figures met with—influence of food—variation with age—variation with kind of animal 109

THE LIVER IN RELATION TO CONSTITUENT SUGAR. Procedure necessary for ascertaining the nature of the sugar. Rapid *post-mortem* production of sugar in the liver. Prevention of *post-mortem* change by boiling and by freezing. Amount and nature of sugar found in the liver respectively before and after the occurrence of *post-mortem* change. Influence exerted by the body-temperature existing at the time of death. State found in the liver of cold-blooded animals. The amount of sugar present in the liver corresponds with that met with in the other structures of the body. The sugar-forming ferment under an inhibitory influence during life. The inhibitory influence removed under certain abnormal conditions. Analogy of the conditions attaching to the coagulation of the blood. Production of sugar in a liver removed and washed free from blood. The ferment possesses a glucose-forming capacity. Production of sugar in liver-substance previously coagulated by alcohol. Proof that the change is not due to living action. Production of sugar in alcohol-coagulated liver-substance increased in presence of blood. Influence of sodium carbonate and citric acid on ferment change in the liver..... 132

THE BLOOD IN RELATION TO SUGAR. Nature of sugar present in blood—glucose. Portal blood an exception. Amount of sugar present in blood—varying statements of different observers. Special precautions necessary in the collection of the blood for analysis. The condition of the urine an index of that of the blood. Arterial and venous blood in relation to sugar—alleged difference—source of the fallacy—precautions required in experimenting to escape from it. Results given.

	PAGE
Question of the disappearance of sugar from drawn blood—Bernard's statement—my own observations at variance therewith. Lépine's Theory regarding glycolysis in blood	157

THE URINE IN RELATION TO SUGAR. Presence of sugar in healthy urine. Causes tending to its concealment. Application of Brücke's process to its separation. Proof of its presence afforded by fermentation and other tests. Photographic representations of osazone crystals obtained from the sugar of the urine of various animals and of the human subject. Amount of sugar in healthy urine. Relation of the sugar of the urine to that of the blood. No tolerating capacity of the blood for obviating escape of sugar with the urine. Proof afforded by sugar-injection experiments. Sugar in urine in proportion to that present in blood ... 178

MUSCLE IN RELATION TO SUGAR. Mammalian animals. Nature of sugar. Photographic representations of osazone crystals from sugar of muscle. Amount of sugar in different muscles and in different animals. Question of the influence of food. Non-mammalian animals 194

THE SPLEEN IN RELATION TO SUGAR. Amount and nature of sugar. Photographic representation of its osazone crystals 200

THE KIDNEY IN RELATION TO SUGAR. Amount and nature of sugar. Photographic representation of its osazone crystals 201

THE PANCREAS IN RELATION TO SUGAR. Amount and nature of sugar..... 202

THE LUNG IN RELATION TO SUGAR. Amount and nature of sugar. Photographic representation of its osazone crystals 202

THE BRAIN IN RELATION TO SUGAR. Amount and nature of sugar........ 204

THE PLACENTA AND FŒTUS IN RELATION TO SUGAR. Amount and nature of sugar.. 204

ORGANS OF GENERATION OF FISH AND CRUSTACEA IN RELATION TO SUGAR. Amount and nature of sugar 205

THE EGG IN RELATION TO SUGAR. Amount and nature of sugar. Photographic representation of its osazone crystals. Change accompanying incubation 206

GLYCOGEN AND PROTEID-CLEAVAGE CARBOHYDRATE. Carbohydrate matter other than sugar belonging to animal system embraced under heading. Bernard's discovery of glycogen. Author's discovery of proteid-cleavage carbohydrate. The terms "glycogen" and "amyloid substance." Glycogen analyses include proteid-cleavage carbohydrate. Comprehension of the two under the term amylose carbohydrate 211

BLOOD AND VARIOUS STRUCTURES OF THE BODY IN RELATION TO AMYLOSE CARBOHYDRATE (PROTEID-CLEAVAGE CARBOHYDRATE AND GLYCOGEN). Blood—difference between portal blood and blood generally. Result of injecting sugar into circulation. Muscle—different muscles and muscle in different animals. Pancreas. Spleen. Kidney. Brain. Lung. Intestinal mucous membrane. Placenta and Fœtus. Generative structures of oviparous animals.. 215

AUTHOR'S CONCLUSIONS.—Glycogenic doctrine to be abandoned. Question, through standing at foundation of application of carbohydrates within the system and pathology of diabetes, one of great importance. Advantage of studying the matter through physiology and pathology conjoined. Review of facts showing that tenets of glycogenic doctrine rest upon a false foundation. Liver, instead of forming sugar prevents its entry into the general circulation, and thus secures escape from diabetes. Representation of conditions existing in health and diabetes. Diabetes a failure or an impairment of the power of disposing of carbohydrate matter before the general circulation is reached. Different degrees of impairment of power associated with different forms of severity of the complaint. A class of case in which defective assimilative action alone exists. Another class of case in which liberation of sugar from the tissues in addition takes place. Rationale of dietetic management in the complaint. Purposes to which carbohydrate matter is in reality applied in the economy of life. Review of what takes place in living nature shows that through the influence of living protoplasm carbohydrate matter undergoes: (1) transmutation; (2) application to the production of proteid; (3) transformation into fat. Universality of action traceable without distinction in the animal and vegetable kingdoms. The transmutation effected by protoplasmic agency is in the direction of dehydration, whilst by ferment action and chemical agents it is in the direction of increased hydration. Illustrations of application of carbohydrate matter through transmutation by protoplasmic agency in the vegetable kingdom. Production of glycogen an instance in the animal kingdom. In plant and animal alike, ferment action first hydrolyses and renders the condition suitable for the exercise of protoplasmic action. Experiments showing that glucose introduced into the circulation gives sugar in the urine with a lower cupric oxide-reducing power. Application of carbohydrate to the production of proteid. Proof afforded by yeast cell. Phenomena observable in the higher vegetable organism. Support derived from the glucoside constitution of proteid matter. Bearings of peptone harmonise with its utilisation in conjunction with carbohydrate in the production of proteid matter. The protoplasm of the cells of the villi the operative synthetic agent. The supply of carbohydrate for proteid formation in part derived from preliminary cleavage, by ferment action of digestion, from the proteids of food. The carbohydrate matter escaping application by the protoplasm of the cells of the villi reaches the portal blood and becomes applied by the supplementary action of the protoplasm of the cells of the liver. Liebig's classification of food principles with respect to the carbohydrates no longer tenable. Transformation of carbohydrate into fat. If carbohydrate matter, as is believed, represents the initial condition of all organic products the whole of the fat of both living kingdoms must take origin directly or indirectly from carbohydrate. Protoplasm a necessary

	PAGE

factor. Carbon may enter proteid molecule as carbohydrate, and come out as fat. Yeast cells afford an illustration of production of fat from carbohydrate. Starch precedes fat in the oily seeds. Growth of oily seeds illustrate the converse process of conversion of fat into carbohydrate. Production of fat from carbohydrate in the animal system. Where and how the process occurs. Lacteals injected with milky chyle, and cells of villi charged with fat after carbohydrate, as well as after fatty, food. Grounds for looking upon the protoplasm of the cells as exerting a transformative action. The cells of the liver perform a supplementary action. Photo-engravings of the cells of the villi, and of the villi in section, after carbohydrate food and during fasting. So-called fatty degeneration of the liver in reality a functional production of fat from carbohydrate. Carbohydrate possibly first incorporated into proteid and fat, subsequently cleaved off. Circumstances suggestive of the fat, and likewise the lactin and casein, of milk being cleavage products from proteid. The fat of adipose tissue may similarly take origin through the intervention of proteid, and myxoedema possibly arise from an imperfect performance of the process. Fatty degeneration of muscle and other tissues the result of conditions favouring cleavage of fat from proteid. Question broadly discussed. A delicately-adjusted balance of the opposing effects of ferment and protoplasmic actions at the foundation of the play of changes belonging to life. The vascular system controlled by nerve influence, and the state of the blood, considered in relation to fat deposition and diabetes.................................. 221

INTRODUCTORY CONSIDERATIONS.

The carbohydrates constitute a sharply-defined group of principles occurring widely and largely in the realm of organic nature.

They are composed of the three elements, carbon, hydrogen, and oxygen. The number of carbon atoms in the molecule is ordinarily six, or a multiple of six,* and the hydrogen and oxygen are always present in the proportion to form water.

The general formula for the group may be represented as $\{C_6(H_2O)_n\}_{n'}$; and, looked at simply from their elementary composition, the constituent members stand throughout in the position of hydrates of carbon. In speaking of them, however, as carbohydrates, it is not considered that there are grounds for regarding them as actual compounds of carbon with water, and the expressions, therefore, that will be subsequently used with reference to transformation, attended with increased or decreased hydration, merely imply an increase or decrease in the proportionate amount of the elements of water.

With this understanding, the term carbohydrates is a convenient one for associating the bodies of the group together and distinguishing them from other non-nitrogenous principles.

In common with other organic compounds, the carbohydrates have their source, directly or indirectly, in the vegetable kingdom. Under the influence of the sun's rays, vegetable protoplasm containing

---

* By artificial means, sugars with 7, 8, and 9 carbon atoms have been constructed from the ordinary 6-carbon-atom sugar. Examples, as my friend Professor Odling, of Oxford, has pointed out to me, of less than 6-carbon-atom sugars, placed in progressive order of descent, are presented by arabinose ($C_5H_{10}O_5$), obtainable by hydrolysis from gum arabic; erythrose ($C_4H_8O_4$), yielded by the oxidation of erythrite, a principle extracted from lichens; glycorose ($C_3H_6O_3$), furnished by the oxidation of glycerine; and glycolose ($C_2H_4O_2$), derived from the oxidation of glycol.

chlorophyll is endowed with the power of dissociating the elements of inorganic principles, and recombining them into organic compounds. It is the great province of the vegetable kingdom thus to play a constructive part in the economy of nature. The living protoplasm of the plant, by virtue of its capability of acting upon matter brought within the sphere of its metabolic influence, serves as an instrument, through the medium of which the solar energy effects the changes that occur; and the energy so expended becomes locked up in a latent, or potential, state in the compound formed, ready to be liberated as actual or kinetic energy when the compound is destroyed. In the plant the destruction of organic matter with its attendant liberation of energy takes place normally to a comparatively insignificant extent, but in certain special operations, as, for instance, notably in flowering and germination, it somewhat more largely occurs. It is pre-eminently in the system of the animal that organic matter is actively destroyed, the energy set free becoming manifest in the varied activities which constitute the characteristic attribute of animal life.

We cannot satisfactorily trace the steps through which the carbohydrates are synthetically formed; but attempts have been made in this direction, one of which, the formic aldehyde hypothesis, may be mentioned by way of illustration. This view assumes that formic aldehyde (CH_2O) is first produced, according to the equation

$$CO_2 + H_2O = CH_2O + O_2,$$

and that this body then gives rise, by polymerisation, or the union of several molecules to form one larger molecule, to a carbohydrate of the composition $C_6H_{12}O_6$, thus—

$$6(CH_2O) = C_6H_{12}O_6.$$

But whatever the nature of the steps, we can say with complete confidence that as the ultimate result carbon is separated from the carbonic acid of the atmosphere, and associated with the elements of water, yielding as the product a carbohydrate.

Instead of the process, however, being a purely chemical one, of the simple nature depicted above, it may happen that the carbohydrate is the outcome of a complex physiological change. It may be that a more or less elementary compound is first formed, which becomes embodied in the substance of the living protoplasm, and that from this the carbohydrate is subsequently thrown off as a cleavage product of protoplasmic activity. To amplify this statement: instead of the carbohydrate being directly, or by intermediate steps, synthesised from its elements, it may happen that these become appropriated by the living protoplasm and worked up into the proteid matter of which protoplasm is constituted, and from which the carbohydrate by a further action is afterwards split off. The *modus operandi* may not be the same, but the splitting off of a carbohydydrate from a nitrogenous molecule is an event that, as is well known, may be brought about by the action of chemical agents and ferments in the case of glucosides.

Pasteur, many years back, showed, as will be more fully considered at pp. 20—21, that a few yeast cells placed in a medium containing tartrate of ammonia, sugar, and the ash of yeast live and multiply. From these simple materials the proteid of protoplasm is thus built up by the agency of pre-existing protoplasm. Further, carbohydrate matter is met with in the yeast organism in the form of cellulose and glycogen: that is, in a lower state of hydration than the carbohydrate matter existing in the pabulum. Seeing, as it must be considered, that the sugar of the pabulum contributes with the ammonia to the production of proteid, it may happen by a subsequent operation that the carbohydrate is split off in an altered form, thus occasioning the transmutation that is observed to occur. The step from the position existing in the case of the yeast cell to that of the protoplasmic matter concerned in the primary formation of carbohydrate matter is not a wide one. The only essential difference discernible is that the proteid matter is constructed by the latter from the more elementary materials carbonic acid, water, and ammonia, in place of the formed carbohydrate and tartrate of ammonia.

I have spoken of the primary formation of carbohydrate matter, and the view taken about its being a cleavage product from proteid matter harmonises with the view that is held with regard to the deposition of starch and cellulose derived from the preformed carbohydrate of the plant. For growth and storage of reserve material, large quantities of carbohydrate matter are being deposited as cellulose and starch. These are derived from carbohydrate matter conveyed in the form of soluble sugar to the seats of deposition, and in the transmutation that occurs it is considered probable that the sugar becomes incorporated with the protoplasmic matter, through the agency of which the operation is performed, and from which the carbohydrate is split off in an altered form. Alike in the primary formation of starch from inorganic materials, and in its secondary formation from pre-existing carbohydrate, the protoplasmic material in which the starch makes its appearance is observed to become diminished in quantity, or, as it were, used up.

The actual seat of primary formation of carbohydrate matter is in the chlorophyll corpuscles. These are little protoplasmic bodies which, under the influence of light, are endowed with the power of appropriating materials derived from the inorganic kingdom to the building up of organic matter.

The form in which carbohydrate matter primarily becomes manifest is as starch. Whether, as is by some surmised, this is preceded by sugar, or whether it is not, the first *visible or demonstrable* carbohydrate product is starch. As the operations of life in the chlorophyll corpuscles proceed, starch granules make their appearance. Sachs, describing the changes perceptible in chlorophyll corpuscles by the aid of the microscope,* says:—" The old observations of Naegeli and myself show that in the primitively quite homogeneous green substance, starch grains, at first extremely small, become visible, usually distributed in twos, threes, or more in the mass of chlorophyll of the corpuscle. These enlarge and, as they meet one another during growth, become flattened and applied close to one another with plane surfaces, while their free sides remain rounded and become arranged more or less according to the form of the chlorophyll corpuscle; occasionally, however, when they arise at the circumference, they protrude from the chlorophyll corpuscle. I also observed almost 20 years ago that, under certain circumstances, when leaves turn yellow without being diseased, the starch grains grow so vigorously in the chlorophyll that the latter becomes, so to speak, entirely displaced by them; and finally, in place of the chlorophyll corpuscle, there lies a starch-grain compounded of several grains."

* 'Lectures on the Physiology of Plants,' by Julius von Sachs, translated by H. Marshall Ward, p. 315: Clarendon Press, Oxford, 1887.

Chemical Constitution.

According to its percentage composition, carbohydrate matter is constituted of carbon with the elements of water, but its behaviour under different conditions and its derivation products may be taken as showing that its molecular construction is of a much more complex nature. I have spoken of its origin as a product of the vegetable kingdom. It is not exclusively, however, through the operations of living matter that its formation is capable of being effected.

Amongst the achievements of modern chemistry has been the artificial synthesis of several carbohydrate bodies, and the discovery of the means of effecting the similar formation of others is probably only a matter of time. The labours of Emil Fischer led some years back to the synthetic formation of a fermentable sugar having the composition of dextrose ($C_6H_{12}O_6$) and closely resembling it in properties, but proving to be optically inactive. This body, designated α-acrose, became the starting point for attaining the synthesis of further products, amongst which were several fermentable sugars of the glucose group, including ordinary dextrose and lævulose.

A further achievement has been the successive production from a six-carbon-atom sugar of sugars possessing seven, eight, and nine carbon atoms in the molecule. The nine-carbon-atom sugar has even been found to be susceptible of undergoing fermentation in contact with yeast. In connexion with these higher carbon-atom sugars the question has been raised as to what might be the effect of feeding animals on them, and the suggestion has been made that possibly modified forms of glycogen, lactose, fat, and proteid might result.

Chemical Characters and Relations.

Of the several carbohydrates known, only a certain number fall within the range of consideration contemplated by this work. Such are cellulose, starch and its congener in the animal kingdom glycogen, the dextrins, maltose, cane sugar, lactose, and the glucoses.

The carbohydrates comprise a class of bodies in which carbon is associated with different proportions of the elements of water, and from the relation existing between the latter and the former, a convenient basis of classification is afforded. Thus regarded, they fall into three groups:—

1. *The amyloses*, with the formula $(C_6H_{10}O_5)_n$, which present the lowest degree of hydration, and stand, therefore, as the initial term of the series. They include cellulose, starch, glycogen, and dextrin;
2. *The saccharoses*, represented by the formula $C_{12}H_{22}O_{11}$, and including maltose, lactose, and cane sugar; and
3. *The glucoses*, which present the highest degree of hydration and possess the formula $C_6H_{12}O_6$. They include dextrose, lævulose, and galactose.

I will proceed to give the main characters and relations, regarded from a physiological point of view, of the above-named bodies arranged under the three specified heads.

The Amyloses.

Cellulose, $(C_6H_{10}O_5)_n$, stands amongst the members of the carbohydrate group that have the least proportion of the elements of water entering into their constitution. It belongs mainly, but not quite exclusively, to the vegetable kingdom, and, constituting as it does the basis material of vegetable cell walls and of woody fibre, it probably ranks as the most abundant organic principle in nature. It takes origin from the soluble carbohydrates or sugars by a process of dehydration effected by the agency of the metabolic power existing in living protoplasm. With the chemical transformation occurring in the production of cellulose through the instrumentality of metabolic action, the exercise of something akin to, if not actually consisting of, plastic or formative power is brought into play in such a manner as to give to the newly formed material a definite structural character. Thus circumstanced, cellulose may be spoken of as taking a place nearer to the position of organised matter than is held by any other member of the carbohydrate group.

Insolubility in the ordinary solvents is one of the chief characteristics of cellulose, but the various forms of cellulose differ in this respect, some being less resistant to solvent action than others. Acids, unless in a concentrated form, do not dissolve it, nor is it dissolved by potash. A ready solvent, however, happens to be afford by an ammoniacal solution of cupric oxide, and from this menstruu. it is precipitable in a flocculent form by acids.

Cellulose gives no colour reaction with iodine alone, but with iodine in the presence of sulphuric acid, or, better, zinc chloride, a blue or violet colour is produced.

The effect of sulphuric acid upon cellulose may be studied by experimenting with cotton wadding. Immersed in the concentrated acid, cotton wadding becomes dissolved, and if the solution be at once largely diluted with water the cellulose is precipitated apparently in an unaltered or but slightly altered form. If, on the other hand, the cellulose is allowed to remain for some time in contact with the acid, dilution with water no longer leads to the production of a precipitate. The cellulose has apparently become transformed into a dextrin-like material, and the solution is found to possess a slight cupric oxide reducing power. Thus transformed, it may be carried further by a strength of acid that has scarcely any effect on untransformed cellulose. This is shown by the greatly increased cupric oxide reducing power that may be produced by boiling after diluting the acid solution to a 2 per cent. strength, under the action of which cellulose is only to a very slight extent influenced.

The resistance offered by cellulose to conversion into a cupric oxide reducing carbohydrate by boiling with 2 per cent. sulphuric acid distinguishes it from starch and glycogen, both of which bodies, on boiling for an hour and a half with the strength of acid named, are completely transformed into glucose. For effecting the conversion, to any notable extent, of cellulose by boiling with dilute sulphuric acid, a 10 or 15 per cent. strength must be employed. At higher temperatures, applied by means of the autoclave, cellulose is attacked by 2 per cent. acid in a manner that it is not on boiling with the inverted condenser, a fact that it is necessary to bear in mind, as will be subsequently explained (p. 63), in the quantitative determination of starch and glycogen where accidental contamination with cellulose from a filter may have occurred.

The component molecules of the composite cellulose molecule seem, to judge from its stability or power of resisting the influence of different agents, to be held together by a tighter bond of union than in the case of other carbohydrates.

Doubtless there is much yet to be learnt about the modifications of cellulose. Even chemically there are differences, some forms being susceptible of undergoing conversion into sugar by an acid of insufficient strength to exert any decided amount of influence upon cotton cellulose, to which the remarks in the preceding paragraph about the effect of acid apply, whilst, physiologically, some of them seem, instead of existing in an independent state, to be more or less bound up with living matter, if not, indeed, in some phase of their history, actually incorporated with it.

Starch, $(C_6H_{10}O_5)_n$, looked at in relation to animal alimentation, may be regarded as by far the most important of the bodies of the whole carbohydrate group.

It is very widely and largely found in the vegetable world, and is the first visible product of that constructive metabolism already described as taking place in all green plants. It is primarily deposited in the leaf, and secondarily, at subsequent stages in its history, in other parts of the plant, and invariably in the form of a granule. The starch granule varies much in size and appearance, according to the particular plant in which it is formed; but it has always a definite structure, consisting of a nucleus or hilum and concentric, or rather, strictly speaking, excentric, layers. This is best made out after treatment with chromic acid or dilute alkali. The granule is not homogeneous, but consists of two isomeric substances, starch granulose and starch cellulose, with, perhaps, intermediate forms. The granulose is much the more soluble, and can be dissolved out by dilute acids, or by boiling, &c. If treated in the cold, the cellulose skeleton remains, retaining the form of the grain, though originally constituting only from 2 to 6 per cent. of the weight of substance. The grain, when intact, is unacted upon by cold water, by reason of the density of its outer layer, but when it is crushed or ruptured the granulose is slowly dissolved out. Under the action of boiling potash the cellulose framework becomes likewise dissolved. The solubility of starch in hot water is to be accounted for by the bursting of the swollen granules, which usually occurs at a temperature of from 50° to 70° C., liberating the granulose and forming the "starch paste."

A solution of starch is a semi-transparent, viscid liquid, which, under examination, exhibits the following properties. In presence of free iodine it yields a characteristic blue colour, which disappears on heating, and reappears on cooling. This behaviour is absolutely distinctive of starch, and affords an easy, delicate, and reliable test for its detection.

Starch solution possesses in a marked degree the property of optical activity, rotating the ray of polarised light strongly to the right. This rotatory power is possessed to a variable extent by most of the members of the carbohydrate group, and on it is based one method of determination applicable to these bodies.

As a colloid, starch does not diffuse through animal membranes. It is absolutely insoluble in alcohol and ether, and is therefore precipitable from its aqueous solution by these agents. It is unaltered by treatment with dilute potash or soda, but boiling with dilute mineral acids carries it into glucose (dextrose) through intermediate cupric oxide reducing products. Diastase and other amylolytic ferments exert an analogous action on starch, except that the process of transformation, in the main, stops short at the stage of maltose, instead of advancing to glucose. There are, however, some ferments in both the animal and the plant which have the power of effecting complete hydration into glucose. Starch is also converted into dextrin by simply heating to between the temperatures of 100° and 200° C.

Starch, like the other amyloses but unlike most of the other carbohydrates, is devoid of taste. It has no cupric oxide reducing power.

Glycogen, $(C_6H_{10}O_5)_n$, may be regarded as the representative in the animal kingdom of the starch belonging to the plant. I consider that nomenclature has been unhappy in the name that has been applied to this body. The term was adopted on the discovery of the material from which the sugar developed in the liver after death takes origin. All the amyloses are, strictly speaking, *glyco-gens;* and, as it happens that the physiological grounds upon which the name was given to the body under consideration prove to be untenable, it is so far a misnomer. *Zoamylin* would be the most appropriate of the hitherto suggested designations to apply to it. Its distribution and relations in the animal organism will be referred to in detail in a subsequent part of this work, and in connexion therewith it will here be sufficient to say that, although constituting a principle especially belonging to the animal kingdom, yet it is not absolutely confined to it. In a large number of fungi it has been found to be present, and it seems in these organisms to take the place of the starch of higher members of the vegetable kingdom. An interesting point is here presented, inasmuch as the fungi constitute a group of organisms differing from the members of the vegetable kingdom generally and agreeing with those of the animal kingdom in the manner in which their aliment is supplied. With the non-existence of chlorophyll corpuscles they present an absence of starch-forming capacity, and, as a corollary, instead of having the power of constructing their living matter from principles of the inorganic kingdom, they are dependent, like animal organisms, for their aliment upon the supply of preformed organic matter. Yeast, it has been discovered, contains glycogen in large quantity, and from this source it may be readily procured and ascertained to exhibit all its characteristic reactions.

Glycogen is soluble in water, but much more so in hot water than cold. It yields with water an opalescent or white milky liquid which perhaps is not, strictly speaking, a true solution. On concentration it assumes a transparent state, and becomes milky again on dilution.

In its various properties and modes of behaviour, glycogen is almost identical with starch, the only noteworthy difference being in the colour produced by treatment with iodine, which in the case of glycogen is a port-wine red, instead of blue as given by starch. The effects produced on the colour by heating and subsequently cooling the solution are the same in each case.

By means of microchemical examination with the employment of iodine, the existence of glycogen in a granular state or in the form of amorphous masses within the liver cells can be demonstrated.

Animal gum (Landwehr), which is to a certain extent allied to glycogen, will receive notice under the consideration of glucosides.

Dextrin, $(C_6H_{10}O_5)_n$, is produced by the action of mineral acids and by diastatic or amylolytic ferments upon the amyloses which have been already described. The transformations effected by the diastatic ferment of malt on starch have been the subject of much careful investigation and of a considerable amount of controversy, and even now it cannot perhaps be definitely said that a full knowledge of the steps of the process has been attained.

The researches of modern investigators point to the occurrence of changes of a highly complex character, and the view at the present time entertained is that from the composite starch molecule molecules become successively split off, and with the splitting off undergo hydration into maltose, leaving at each stage a diminished dextrin molecule of unaltered percentage composition. Thus each successive stage in the process yields simultaneously a substance of higher hydration and a dextrin residue, which does not differ in percentage composition from the original starch molecule. The hydrated portion split off at each step is, according to Brown and Morris, either maltose ($C_{12}H_{22}O_{11}$) or an intermediate product [either maltodextrin $(C_{12}H_{20}O_{10})_2C_{12}H_{22}O_{11}$ or amylodextrin $(C_{12}H_{20}O_{10})_6C_{12}H_{22}O_{11}$], which speedily undergoes further hydration into maltose.

The successive unhydrated residues have all been termed dextrins. Whilst retaining the same percentage composition, they present a steadily diminishing molecular weight: they are all, that is to say, represented by the formula $(C_6H_{10}O_5)_n$, n being diminished at each successive stage of splitting off.

By the continued action of the ferment the main portion of the starch undergoes conversion through intermediate stages of dextrin into maltose. There remains, however, a portion which offers greater resistance than the rest to the change, and which is with difficulty made to undergo hydration. This is the dextrin which is most easily isolated and obtained for examination, and to which the descriptions given of dextrin more especially apply. According to the latest researches of Brown and Morris, it constitutes a fifth of the original molecule of soluble starch, the view founded upon these researches being that the first effect of the diastatic ferment is to lead to the constituents of the molecule becoming ranged into five separated groups, one of which is so constituted as to be more stable than the others, by virtue of which it remains intact after the others have undergone hydrolysis.

The dextrins which are first formed in the process of hydrolysis are coloured red by iodine, and are on this account distinguished as *erythrodextrins;* whilst those produced from them and met with later yield no colour with iodine, and are called *achroodextrins*.

Dextrin is produced, not only by the action of ferments and acids upon starch or glycogen, but also by the mere heating of these bodies when the temperature approaches 200° C.

Dextrin is uncrystallisable, and when dried is a glassy colourless body susceptible of being ground down to a white powder. It is easily soluble in water, forming a clear solution possessing no decided taste. It is precipitated by strong alcohol. Like starch and glycogen, it is dextrorotatory, whence its name. Authorities have been somewhat at variance as to whether any of the dextrins are possessed of cupric oxide reducing power. They are, however, generally considered to have no reducing power.

The Saccharoses.

The principles falling under this head requiring notice are maltose lactose, and cane sugar.

Maltose ($C_{12}H_{22}O_{11}$), arising from the hydration of starch by the incorporation of water, thus, $2C_6H_{10}O_5 + H_2O = C_{12}H_{22}O_{11}$, is a body soluble in water, alcohol, and ether, and readily diffusible. It crystallises in the form of hard, white, fine needles. Like the amyloses, it is optically active and dextrorotatory, though in a lower degree; but, unlike them, it has a distinct taste, which is faintly sweet. It is, moreover, unaffected by iodine, as indeed are all the carbohydrates of higher hydration than the amyloses, and it has a cupric oxide reducing power equivalent to 61 as compared with that of glucose taken at 100.

Although maltose constitutes the end product of the action of diastase, and amylolytic ferments generally, on starch and its congeners, it is susceptible of being carried on into glucose (dextrose) by boiling with dilute mineral acids, as also by the agency of certain ferments. The following expression represents the change:—

$$C_{12}H_{22}O_{11} + H_2O = C_6H_{12}O_6 + C_6H_{12}O_6.$$
Maltose.   Dextrose.   Dextrose.

Saccharose or *Cane Sugar.*—Identical in composition with maltose, though without any genetic relation to the amylose bodies, is saccharose, sucrose, or cane sugar ($C_{12}H_{22}O_{11}$). This sugar is found in parts of many plants. It is a substance very soluble in water, and is also, though less easily, dissolved by alcohol. It crystallises in large monoclinic prisms, and as a crystalloid is readily diffusible. It does not reduce cupric oxide.

A solution of cane sugar is, like all the bodies hitherto described, dextrorotatory. By prolonged boiling, however, with water, or by boiling for a short time with dilute acid, or by the action of yeast and of certain unorganised ferments at ordinary temperatures, it undergoes a change with increase of hydration, the first indication of which is the reversal of its effect on polarised light. It now rotates the ray to the left instead of to the right, and from this fact has received the name of invert sugar. This invert sugar on examination proves to be a mixture of the two glucoses dextrose and lævulose in equal proportions. The greater optical activity of lævulose accounts for the lævorotatory power possessed by invert sugar. The inversion of cane sugar presents another instance of transition from the saccharose to the glucose group, and evidently consists in a simultaneous hydration and decomposition of the molecule, according to the equation

$$\underset{\text{Cane sugar.}}{C_{12}H_{22}O_{11}} + H_2O = \underset{\underbrace{\text{Dextrose.} \quad \text{Lævulose.}}_{\text{Invert sugar.}}}{C_6H_{12}O_6 + C_6H_{12}O_6.}$$

Lactose or *milk sugar* ($C_{12}H_{22}O_{11}$) is only known to occur in the animal kingdom. It constitutes the saccharine principle belonging to milk. It has a faintly sweet taste. It is soluble in water, but much less so than cane sugar. It is insoluble in alcohol and ether. Milk sugar crystallises in white, rhombic prisms. Its solution is dextrorotatory, and exhibits the phenomenon of birotation : that is to say, the freshly made solution causes a rotation twice as great as that which it will produce after standing.

Lactose possesses cupric oxide reducing power, but discordancy exists in the precise figures given for it. Many authorities express the power at 74, as compared with that of glucose taken at 100. Figures, on the other hand, as low as 52 have been assigned to it. Unlike glucose, lactose does not effect a *ready* reduction of the copper test solution, and thus the terminal point is modified according to the precise manner in which the examination may happen to be conducted. I have made a large number of observations with the employment of the ammoniated cupric liquid. As in the case of Fehling's solution, the terminal point of reduction is less sharply defined than when other sugars are employed, but the result of my observations is to place the cupric oxide reducing power of lactose at 60, or a little over.

It seems, judging from what I have observed, that under boiling with acetic and citric acids lactose undergoes a certain amount of modification without being converted into glucose. Boiling for from half an hour to an hour with citric acid of from 2 to 5 per cent. strength has the effect, according to the results before me, of raising the cupric oxide reducing power to about 74, beyond which it cannot be carried by further boiling. After boiling with a similar strength of acetic acid, a reducing equivalent of about 64 or 65 has been found to be given.

Harmonising with these differences revealed through cupric oxide reducing action, differences are noticeable in the osazones produced with phenyl-hydrazine. Lactose in its unaltered form yields, I find, contrary to what is generally stated, a deposit of an amorphous, or non-crystalline, character, presenting the appearance of minute spores or beads. Such is what is noticed when, for example, the crystallised lactose of commerce is taken and treated with phenyl-hydrazine hydrochloride and acetate of soda, or phenyl-hydrazine with acetic acid in the amount ordinarily recommended for employment. If, however, the lactose has been previously boiled with, say, 5 per cent. acetic acid, or even if the acetic acid used in the application of the test is added in considerable excess, a crystalline osazone is produced, presenting the appearance of irregularly curved, whip-like filaments, radiating from a central nucleus. After boiling with 5 per cent. citric acid, and, whether acetic acid is subsequently added in conjunction with the phenyl-hydrazine or not, a crystalline osazone is given, in which the wavy filaments from the nuclear masses are replaced by straight spines, or flat blade- or lancet-like projections. After boiling with sulphuric acid, the long straight needles and radiate clusters of dextrosazone and galactosazone are presented.

In experimenting upon lactose derived direct from milk, similar results are obtained, but it is necessary to guard against being misled by the process adopted for separating the casein and fat preparatory to the application of the test. If the separation be made by rennet, rendered neutral to test-paper, the amorphous deposit is given with phenyl-hydrazine hydrochloride and acetate of soda, or with phenyl-hydrazine and acetic acid, unless the latter is used in excess. If, on the other hand, the separation be made by acidification with acetic acid and the application of heat, the crystalline osazone described above as yielded by lactose modified by acetic acid is given.

These results are instructive, by bringing into view constitutional modifications that were not previously known to exist, and suggestive that much of a like nature remains to be discovered with respect to other bodies.

Boiling with dilute sulphuric acid converts lactose into a mixture of the two glucoses dextrose and galactose in equal parts, thus:—

$$\underset{\text{Lactose.}}{C_{12}H_{22}O_{11}} + H_2O = \underset{\text{Dextrose.}}{C_6H_{12}O_6} + \underset{\text{Galactose.}}{C_6H_{12}O_6}.$$

The glucose thus formed gives with phenyl-hydrazine, as already mentioned, the characteristic dextrosazone needles, intermixed with radiate clusters of acicular crystals—galactosazone.

Milk sugar readily undergoes the lactic acid fermentation, and various micro-organisms have been mentioned as productive of the change.

The Glucoses.

The only members of this group which need be here referred to are dextrose, lævulose, and galactose. All have the composition $C_6H_{12}O_6$.

Dextrose, known as *grape sugar*, is to the animal physiologist the most important of the glucoses. It occurs widely in the vegetable kingdom, usually in company with lævulose. The two are found in most sweet fruits, and also in honey. Dextrose is the form of sugar that occurs in the urine of diabetic subjects. It is best obtained on a large scale by the action of dilute sulphuric acid on starch; and, by similar means, it is also obtainable from the carbohydrates generally of lower hydration than itself. From glucosides it is likewise derivable by the action of acids, and in some cases by that of ferments.

Dextrose dissolves in its own weight of cold water, and is soluble also, though far less readily, in alcohol. It is less sweet to the taste than cane-sugar, and is highly diffusible. It crystallises in microscopic, rhombic plates, aggregated into nodular, warty masses. It has the maximum extent of cupric oxide reducing power possessed by the carbohydrates. Dextrose, subjected to the influence of dilute acids and the unorganised ferments, undergoes no further hydration change. It readily undergoes the alcoholic fermentation in contact with the growing cells of yeast. Dextrose, as its name implies, is dextrorotatory. It exhibits the phenomenon of birotation.

Lævulose is a constituent, along with dextrose, of most sweet fruits, and also of honey. As previously stated, it is a product of the inversion of cane sugar. Its optical activity is greater in amount than, and opposite in kind to, that of dextrose: that is to say, it rotates the ray of polarised light further to the left than dextrose does to the right. Hence the lævorotation exerted by invert sugar.

Lævulose is more soluble than dextrose, both in water and in alcohol. It is also sweeter to the taste, being in fact as sweet as cane sugar. Its power of reducing cupric oxide is the same as that of dextrose. In its chemical relations, lævulose closely resembles dextrose, but it is more susceptible of being altered by heat and acids, and less susceptible to the action of alkalis and ferments. In connexion with the remark that lævulose is more readily acted upon by acids than dextrose, I may mention that in the analytical determination of the nature of a sugar through the cupric oxide reducing power presented before and after boiling with sulphuric acid, it is found that where lævulose is dealt with a certain amount of loss is apt to ensue from the action of the acid, and thus give rise to lower figures being obtained after the boiling than before.

Lævulose crystallises from its alcoholic solution in the form of fine silky needles.

Both dextrose and lævulose, on exposure to a temperature of 170° C., lose water and become converted into bodies of the composition $C_6H_{10}O_5$, called respectively glucosan and lævulosan, both of which are susceptible of reconversion by the agency of mineral acids into their original state.

Galactose is obtained, together with dextrose, by boiling milk sugar with dilute mineral acids. It is dextrorotatory, and exhibits birotation. It possesses cupric oxide reducing power. It is much less soluble in water and is more readily crystallisable than dextrose or lævulose, and differs chemically from these bodies in yielding mucic acid, and not saccharic acid, when oxidised. Otherwise it very much resembles them in its properties.

BEHAVIOUR OF SUGARS WITH PHENYL-HYDRAZINE.

Osazones.

During the past decade the researches of Emil Fischer, founded upon the discovery that combinations of cupric oxide reducing sugars with phenyl-hydrazine exist, have thrown much light on the constitution of the sugars, and have afforded a valuable means of separating and identifying them. It is conceded that the sugars constitute bodies belonging to the aldehydic and ketonic groups, and, as with all aldehydes and ketones, the cupric oxide reducing sugars form compounds with phenyl-hydrazine—compounds which possess definite crystalline characters. The product is formed in two stages. In the first place, one molecule of the sugar becomes united with one molecule of phenyl-hydrazine, giving rise to a *hydrazone*, a body, generally of a soluble nature, which in presence of an excess of phenyl-hydrazine leads on to the formation of a second compound. The hydrazone first formed undergoes oxidation to a certain extent, and combines with a second molecule of the phenyl-hydrazine, producing an *osazone*, a body of slight solubility, and usually of a crystalline nature. The osazones derived from the different sugars present characteristic differences as regards crystalline form, melting point, solubility, and optical properties. By appropriate treatment the sugar may be recovered from the osazone, and, moreover, it has been found possible, through the intervention of the osazone, to convert one sugar into another, as, for instance, dextrose into lævulose.

Of the various osazones, glucosazone stands out as the one that reveals itself most readily and most conspicuously. It, indeed, often separates out shortly after starting exposure on the water-bath, and may subsequently appear in such abundance as to give to the product under examination a solid consistence. With many sugars, the heating on the water-bath requires to be prolonged for from one to two or three hours, and even then the osazone may not separate out till after cooling or, it may be, till after standing for some hours.

TRANSMUTATION OF CARBOHYDRATES.

In the carbohydrates we have a group of bodies presenting, as we have seen, varying degrees of hydration, and it is further noticeable with regard to them that they can be carried from one degree of hydration to another in both directions. By certain means, which, whilst operating within the living organism, can at the same time be thrown into play at our command outside the organism, transmutation by increased hydration can be effected. Passage in the direction of diminished hydration is very largely taking place around us, but, with the trivial exceptions to be subsequently mentioned, it is only, so far as appears from observation, through the agency of the conditions existing in connexion with actually living protoplasmic matter that it is susceptible of being brought about. I will give consideration separately to the two kinds of transmutation.

Transmutation of Carbohydrates by Increased Hydration.

The carbohydrates of lower hydration are easily moved into a state of higher hydration by the action of acids and ferments, and by both these agencies we have the power of effecting at will the transmutation. In this way, for instance, the amyloses may be made to pass into the group of saccharoses, as in the conversion of starch into maltose; and the saccharoses into the glucoses, as in the conversion of maltose into dextrose, and saccharose into invert sugar.

The effect of acids in the transformation of carbohydrate matter is so well known that the subject does not need consideration here.

Ferment action must not be confused with fermentation. Under the head of fermentation are included transformations such as that of sugar into alcohol and carbon dioxide, effected by the growth of the *Torula cerevisiæ* or yeast plant; that of milk sugar into lactic acid, by the various micro-organisms to which the action has been assigned; that of alcohol into acetic acid, by the *Mycoderma aceti*; those of putrefaction by bacteria; and perhaps transformations to the products of which the evil effects produced upon the invaded organism by pathogenic bacilli may be in part attributable.

In these transformations we have to deal with a molecular disruption of organic matter lying outside the living cell organism, but within its sphere of influence—a disruption induced as a collateral effect of the changes of growth taking place within the organism.

Ferment action, on the other hand, is brought about by the agency of unorganised, though organic, material. It consists essentially in a process of hydrolysis—a splitting-up of a composite molecule into segregated parts accompanied with a fixation of the elements of water. As examples may be mentioned, the conversion of starch and other amyloses into maltose by diastase, ptyalin, and the amylolytic ferment of the pancreas; the conversion of cane sugar into invert sugar by invertin; and the conversion of albumin and such-like proteids into peptone by pepsin and trypsin.

Through zymolysis, or ferment action, even bodies of a very stable nature, looked at from a general point of view, are susceptible of being broken down and dissolved. Cellulose, for example, which is resistant to the solvent influence of ordinary chemical agents, may be brought with facility by ferment action into a state of solution.

A striking characteristic of the ferment, or enzyme, which effects the transformation is its power of inducing an indefinite, indeed almost an unlimited, amount of change, without itself undergoing any appreciable alteration or loss. Further, an almost infinitesimal amount of the ferment is sufficient to produce a very extensive effect. In fact, the amount may be so small that the presence of the ferment is only susceptible of recognition through the ferment action to which it gives rise.

In constitution the enzymes appear to be of the nature of proteids, but it cannot be considered as absolutely established that they are so. They are soluble in water, insoluble in absolute alcohol, and non-diffusible. They need not exist at their source of production in a free form, but may become developed from an antecedent zymogen at the moment the suitable conditions happen to be supplied.

Transmutation of Carbohydrates by Decreased Hydration.

Transmutations in the direction of diminished hydration are not under our control to bring about in the same way as those attended with increased hydration, and, as a broad proposition, it may be stated that it is only through the intervention of the power belonging to living matter that they occur.

A few instances, it is true, can be adduced in which, by artificial means, transformations from the higher to the lower forms of hydration can be effected. Thus dextrose ($C_6H_{12}O_6$), when heated to 170° C., loses a molecule of water, and is converted into glucosan ($C_6H_{10}O_5$), a body having the percentage composition of the amyloses. Lævulose ($C_6H_{12}O_6$) also, similarly treated, is converted into lævulosan ($C_6H_{10}O_5$); and cane sugar ($C_{12}H_{22}O_{11}$), heated a little beyond its melting point (160° C.), becomes transformed into a mixture of lævulosan and dextrose, thus

$$C_{12}H_{22}O_{11} = C_6H_{10}O_5 + C_6H_{12}O_6.$$
Cane sugar. Lævulosan. Dextrose.

Both glucosan and lævulosan are reconvertible into glucose by boiling with dilute mineral acids. As another instance, may be mentioned the transformation into dextrin which dextrose is said to undergo (Musculus) when it is dissolved in strong sulphuric acid and subsequently poured into 95 per cent. alcohol.

Some years ago I thought I had obtained evidence of the carrying down of carbohydrates by ferment agency outside the body, but from knowledge since acquired I have discovered that a source of fallacy, which I was not alive to, existed. Indeed, general observation is to the effect that it would be contrary to the usual order of events for ferment action to produce dehydration, its characteristic effect being to produce change in the opposite direction.

There can be no doubt that transmutation by dehydration is very largely taking place in connexion with the operations of life, and examples of its occurrence are readily forthcoming.

One of the most striking and conclusive, on account of the simple nature of the conditions existing, is that which is afforded by what takes place as a result of the growth of yeast. From the researches of Pasteur it is known that a few yeast cells placed in a medium consisting of water, tartrate of ammonia, cane sugar, and mineral matter derived from the ash of yeast, grow and multiply, and in doing so supply evidence, not only of the production of proteid matter from the simple materials named, but also of the dehydration of carbohydrate matter. The carbohydrate in the pabulum is in the form of saccharose, and from this we obtain the products of fermentation, products issuing from a change taking place *outside* the yeast organism, and also cellulose and glycogen (*vide* p. 10, regarding the existence of glycogen in yeast), representations of transformed carbohydrate matter existing *within* the organism. As a concurrent event associated with fermentation, in the case of saccharose, and, in fact, as a preparatory step to it, the saccharose is raised in hydration to glucose by a ferment—invertin—belonging to the yeast, before the splitting up process occurs. Whether in the appropriation also of saccharose within the yeast organism it is previously carried into glucose or not, in either case the production of cellulose and glycogen constitutes an act of dehydration, effected by the protoplasmic matter of the cell.

Instances of dehydration are afforded in the higher vegetable organisms by the production, from the sugar of the sap, of starch, and more rarely of inulin and cellulose, for storage as reserve materials, and also of cellulose for deposition as a textural material in the process of growth.

The production of saccharose from glucose in the ripening of the sugar-cane furnishes another illustration of dehydration. I am indebted for direct information upon this point to one of my former assistants, Mr. Scard, who is now at the head of the chemical department of the Demerara sugar estates of the Colonial Company, and whose duties include the study of the conditions influencing the production of sugar in the cultivation of the cane. It appears that during the period of active growth of the cane the proportion of glucose existing in the juice is far greater than when the cane reaches maturity. As the cane ripens, a gradual diminution of glucose and increase of cane sugar—the one standing in proportion to the other—are observed to take place. During the active period of "arrowing," or flowering, the proportionate amount of glucose is large. On the disappearance of the "arrow," the cane sugar increases at the expense of the glucose, and at the end of about a fortnight the glucose is at its minimum, and the cane sugar at its maximum. Subsequently, with the renewed growth of the cane, attended with the throwing out of new shoots, the glucose again increases. As a further point, it is to be stated that in the upper or growing portion of the cane glucose and cane sugar co-exist in nearly equal proportions, whilst in the lower segments, or more matured part, cane sugar is present in larger amount than elsewhere, and is accompanied with only traces of glucose.

An apposite illustration of the occurrence of dehydration in the animal kingdom is afforded by the production of glycogen in the liver from sugar derived from the carbohydrates of the food and conveyed to it in the portal blood. The matter in question will form the subject of consideration in subsequent pages of this work.

Transmutation of Carbohydrates by Ferment and Protoplasmic Actions, considered in relation to the Operations of Life.

The effects produced by ferment action, on the one hand, and by metabolic protoplasmic action, on the other, are of an opposite nature. In the one case, more or less highly complex molecules of matter become split up into simpler ones with the occurrence of hydration. In the other, constructive and dehydrating operations are carried out. Matter existing in, or which has been brought into, a more or less simple molecular state is influenced in such a manner that combination ensues, and the more complex molecules belonging to the living organism are thus built up. Both operations proceed upon identically the same lines in the two kingdoms—animal and vegetable, of nature.

Ferment action, as has been stated, breaks down complex molecules into simpler ones, and hydrates. Non-soluble and non-diffusible matter—that is, matter the molecules of which are presumably too large to pass through membranous septa—by such agency broken down, becomes soluble and diffusible. With the change effected, it is placed in a position to be susceptible of absorption, and thereafter of transportation from one part of the living organism to another, whereby it is brought within the sphere of influence of protoplasmic matter for appropriation or utilisation in the living economy.

It is by the amylolytic, or diastatic, ferments that carbohydrate matter is acted upon. Proteid matter yields, in a similar way, to the proteolytic, or peptonising, ferments. The effects wrought upon the two kinds of matter by the respective ferments are of the same nature, and the two kinds of ferment action occur in common in animal and vegetable organisms. The transformation of carbohydrate matter, as an event pertaining to life, is easy of observation alike in the plant and the animal. The transformation of proteid matter is also an event pertaining to life that has long been known to be easily susceptible of observation in association with organisms of the animal kingdom. Latterly, proteolytic ferment action has been recognised as occurring in the vegetable kingdom to an extent that was not formerly suspected. It is not improbable that proteolytic ferment action is a phenomenon of universal occurrence in the plant, and instances certainly can be brought forward in which the existence of an active peptonising ferment is susceptible of ready demonstration. The most notable example is found in the *Carica papaya*; and, in reference to this, the remark admits of being made that from papaw juice so large an amount of active peptonising ferment, called *papain*, can be extracted that a place has been given to the plant amongst the *materia medica*. As regards the existence of peptonising ferments in the vegetable kingdom, Sachs* says : "Attention was first drawn to the occurrence of peptonising ferments in the vegetable kingdom by the remarkable phenomena observed in the so-called insectivorous plants. My earlier studies on the germination of various seeds left no doubt that seedlings dissolve and make active their proteinaceous reserve materials by means of peptonising ferments. Gorup-Besanez was, however, the first to detect peptonising ferments in seeds . . . . More recently, a very energetic peptonising ferment in the latex of *Carica papaya* has attracted particular attention, and a similar ferment has been detected in the latex of the common fig (*Ficus carica*). As we come to know the proteinaceous reserve materials of plants better, and if we follow their behaviour in the animal body also, it can scarcely be doubtful that, in spite of incomplete knowledge, the assumption is, nevertheless, warranted that peptonising ferments are perhaps universally distributed in plants; moreover, peptones, the result of their activity, have actually been detected by Schulze in the seedlings of the Lupine."

I have entered into these considerations bearing on proteids on account of their undergoing, under the influence of ferment action, the same kind of change as carbohydrate matter; and, further, on

* 'Lectures on the Physiology of Plants,' by Julius von Sachs, translated by H. Marshall Ward: Clarendon Press, Oxford, 1887, pp. 341—345.

account of the interest attaching to the analogy to be traced in the phenomena occurring in the plant and animal.

Looked at in relation to what has preceded, the province of ferment action is to prepare for the exercise of protoplasmic action. But ferment action probably, also, plays a part in connexion with the multifarious retrogressive changes occurring within the system, from some of which carbohydrate matter may take origin as a product. The subject, however, is not as yet one that can be considered ripe for profitable discussion.

Whilst ferments split up and hydrate, the processes of synthesis and dehydration are the result of the action of living protoplasmic matter. It is a property of living matter to possess the power of converting suitable principles, brought within its sphere of influence, into the likeness of itself. The power is exerted in two directions: in giving form, and in producing chemical change. To these two manifestations of power the terms "plastic" and "metabolic" were, many years ago, applied by Schwann. Through the agency of the "plastic" power possessed by living protoplasm, matter in a previously liquid or amorphous condition acquires determinate or definite form, and, similarly, through the "metabolic" power, undergoes changes of a chemical nature. This metabolic power it is that is at the foundation of the chemistry of living nature—a chemistry which is characterised by its capacity of leading, amongst other results, to the construction of products of more or less complex molecular constitution, a certain number only of which have as yet proved susceptible of being formed by the operation of forces brought to bear, in laboratory undertakings, outside the body. Although it thus happens that products are formed by the agency of living matter which the chemist as yet has not been able to form through the operation of chemical forces in the laboratory, it is not to be contended that a different kind of chemical force exists in living from that existing in non-living matter, but merely that in its operation in connexion with living matter, it is brought into play in association with circumstances of a different nature.

I have spoken of the effects produced by ferment action and protoplasmic action. I will now give illustrations showing how these actions take their turn in the play of changes belonging to life. The illustrations will be drawn from the vegetable kingdom, where the operations of life can be more easily followed than in animal organisms. They will assist in unravelling the more complicated operations of animal life that will come before us for consideration later on in this work.

The primordial, or first-formed, starch, developed in the chlorophyll corpuscle, is a product of protoplasmic action. By ferment action it is transformed into sugar, which passes in the sap to seats where the operations of growth and storage are going on. Here protoplasmic action again comes into play, and through its agency the sugar is reduced in hydration and converted into cellulose or starch, or, it may be, some other allied principle.

Take again the starch which has been deposited in a grain of wheat or other seed, by protoplasmic action, as storage material for service in connexion with the evolution of the embryonic organism. Whilst the formation of the seed is taking place, life exists in its growing structure, and it is through the property of, or the power possessed by, the living matter that the sugar reaching it from the juice of the plant is transmuted into, and deposited as, starch. When the developmental process concerned in the production of the seed is completed, life, except in the small part constituting the embryo, ceases to exist, and transmutation in the direction of that effected by protoplasmic action can no longer occur. The storage matter possesses within itself no power of resuming or re-acquiring the living state which existed in connexion with it at the time of its deposition. But in close proximity to it there lies a minute collection of matter imbued with the power of starting into living activity when the requisite conditions are supplied. This, the embryo, is the part from which the act of germination proceeds, and from it there is developed the ferment which leads to the conversion of the stored starch into sugar. Sachs* says: " The ferments appear to be always produced by the growing parts of the seedlings and buds themselves, and to penetrate from these into the reservoirs of reserve materials, there to dissolve or make active the constructive materials. This is particularly evident in the case of seeds containing endosperm. If the young seedling (embryo) is removed from the seed of the Indian corn (maize), barley, or other plant, and the endosperm alone laid in moist warm earth, its starch is not dissolved and transformed into

* *Loc. cit.*, pp. 343—344.

sugar." Alongside these words of the scientist of the present day, I cannot refrain from placing the words of the poet, written nearly a couple of centuries ago, which not only show the knowledge that at such time had been acquired by observant attention, but, in an eloquent manner, give expression to it. In the subjoined passage* the words to which attention is directed are inserted in italics.

> "Tell me why the *Ant*
> Midst Summer's plenty thinks of Winter's want:
> By constant journeys careful to prepare
> Her stores; *and bringing home the corny ear,*
> *By what instruction does she bite the grain,*
> *Lest hid in Earth, and taking root again,*
> *It might elude the foresight of her care?*"

The sugar which has been produced by ferment action from the stored starch, being conveyed to, and falling within the sphere of influence of, the living protoplasm of the embryo, becomes appropriated and metamorphosed into the cellulose developed and deposited as a constituent of the growing organism.

As a further illustration, I may cite what occurs in the case of the tuber, and take as an example the potato. Starch, in the first instance, is formed and deposited by protoplasmic action from sugar derived from the juice of the plant. With the completion of deposition, the protoplasmic activity which has been previously in operation ceases to manifest itself. In certain parts of the tuber, however, there are little buds—"eyes," as they are called—which, as with the embryo of the seed, under exposure to conditions favourable to growth, burst forth into active life and produce a ferment that acts upon the surrounding starch, reconverting it into sugar. The sugar thus formed is next, by the protoplasmic action of the living matter of the bud, reduced in hydration and transmuted into the cellulose entering into the structure of the growing shoot.

* 'Poems on Several Occasions, by Matthew Prior, Esq.': London, printed for T. Johnson, 1720.

GLUCOSIDES.

The glucosides have long been known to chemists as a class of bodies which, by the agency of ferments and by the action of acids and alkalis, and even, to a slight extent, of water at elevated temperatures, undergo a cleavage or disruption, with sugar as one of the products. Formulæ are given in text-books of chemistry representing the molecular change that occurs, and they show that the phenomenon is usually attended with the incorporation of one or more molecules of water.

The group comprises bodies of very variable composition. In some (salicin may be mentioned as an example) only the three elements carbon, hydrogen, and oxygen are present. In others—as an instance, amygdalin—nitrogen in addition exists. In myronic acid ($C_{10}H_{19}NS_2O_{10}$), a glucoside obtained from the seed of black mustard, there is the further incorporation of sulphur. Another step in the direction of increasing complexity carries us to a body standing in close proximity to the proteids, viz., mucin, which, as a constituent of connective tissue as well as of mucus, exists extensively diffused throughout the animal system. The researches of Landwehr have shown that, under certain treatment, mucin yields a non-reducing carbohydrate, which he has described under the name of "animal gum," possessing the formula $(C_6H_{10}O_5)_n$, and that this is convertible into a cupric oxide reducing, but non-fermentable, sugar, having the composition of glucose ($C_6H_{12}O_6$), which he calls "gummose."

The announcement of the glucoside constitution of mucin was at first received with some mistrust, but the view has now sufficiently gained credence to meet with recognition in standard works on physiological chemistry.

My own investigations carry us yet a step further, and bring the extensive group of proteids of both the animal and vegetable kingdoms of nature into the class of glucosides. I will proceed to show how I was led up to this discovery, and upon what grounds the statement I have made is based. I will afterwards discuss the glucoside constitution of proteid matter, looked at in relation to the chemistry of life. In the discussion the utilisation of carbohydrates in the construction of proteid matter, and also the knowledge that has been acquired regarding the nutritive chemical changes that occur in connexion with vegetable life, will receive consideration. From this comprehensive survey a view will be seen to be opened out, giving to glucosides a position of the deepest physiological interest and importance, by bringing them significantly into participation, as intermediary agents, in the play of changes appertaining to the chemistry of life.

The announcement of the discovery of the glucoside constitution of proteid matter was made in a communication read at the Royal Society, June 8th, 1893, of which the subjoined is a transcript.

"THE GLUCOSIDE CONSTITUTION OF PROTEID MATTER."*

"At quite an early period of my research work I adopted a process for separating the glycogen of the liver, which consisted in boiling with potash, pouring into alcohol, and collecting the precipitate. For the purpose of estimation, the precipitated glycogen was converted by means of dilute sulphuric acid into glucose, the determination of which gave the information required. This process I afterwards applied to blood and the various organs and tissues of the body, with the result that, in all cases, a more or less notable amount of cupric oxide reducing product was obtained. This I looked upon as taking origin, as in the case of the liver, from glycogen. I gave particulars of the amounts derived from various sources in a communication presented to the Royal Society in 1881 ('Proceedings,' vol. 32, p. 418).

"In operating upon small quantities of blood, &c., for quantitative analysis, no difficulty was experienced in obtaining what I took to be glycogen, from its being convertible, like the glycogen in the case of the liver, into a cupric oxide reducing product, by the agency of sulphuric acid; and from my analyses I obtained very accordant results. It stood otherwise, however, when large quantities were operated upon with the view of collecting the product for the purpose of studying its characters. In these attempts, which were undertaken

* In abstract, 'Proceedings of the Royal Society,' vol. 54, p. 53.

at different times upon blood, eggs, and the spleen, I invariably failed to obtain anything like the amount that ought to have been yielded according to the indications afforded by the quantitative analysis conducted. It was obvious to me that there was something connected with the extraction with which I was not acquainted. Perplexed at the loss that was encountered, I resolved to push inquiry, and see if the discrepancy could not be cleared up. From what I will proceed to adduce, it will now be seen that the explanation of my former want of success is sufficiently intelligible.

"A start was given by the following discovery:—It chanced that my research assistant, Mr. W. S. Rowntree, conducted some examinations, in one set of which the analysis was uninterruptedly proceeded with, whilst in the other set, duplicate specimens, after being placed in contact with potash, were allowed thus to remain for several days before the subsequent steps of the process were carried out. The figures derived from the latter stood higher than those from the former, and the difference was sufficiently marked to arouse my attention, and lead me to conclude that it could only be due to the effect of the varying exposure to contact with potash.

"Pursuing the suggestion emanating from what had been observed, I instituted a series of experiments, in which various products were exposed to the influence of potash for different lengths of time. It will suffice here to cite the results obtained from those upon muscle, which, from the little colour to be dealt with, yields a very favourable material for experimenting with. The muscle, after having been thoroughly extracted with alcohol, was dried and reduced to a finely-divided state. Equal portions were taken, and placed in contact with equal quantities of a 10 per cent. potash solution. After being allowed to remain for varying periods in this state, they were boiled, poured into alcohol, and afterwards proceeded with in the usual way. The results obtained stood as follows:—

	Cupric oxide reducing power, expressed as glucose per 1000 parts of dried muscle.
Boiled at once with potash	35·6
,, after standing 3 days with potash..	41·8
,, ,, 6 ,, ,, ..	58·1
,, ,, 10 ,, ,, ..	59·2
,, ,, 14 ,, ,, ..	58·1

"After having observed the manner in which the result was influenced by the duration of the exposure to the action of the potash, I tried the effect of altering the strength of the potash solution employed, and took for experiment separated proteids derived from various sources. The observations showed that a marked variation occurred as a result. A 2 per cent. strength, it was found, suffices for dissolving the material and subsequently yielding a good liquid for titration with the ammoniated cupric solution, but the amount of cupric oxide reducing product resulting from its use stands far short of that met with where a 10 per cent. solution has been employed.

"The steps of procedure were these: The material was in each case taken in a water-free state. Preparatory to treatment, it was pulverised in a mortar and passed through a fine metallic gauze sieve (90 to the linear inch). Minuteness of subdivision is an important condition for securing complete solution by the potash. If boiled in a coarse state with potash, some particles may escape solution and disintegration, and thus lead to untransformed proteid matter being subsequently present when the stage of treatment with sulphuric acid is reached, the effect of which is to give a violet or rose-red colour (biuret reaction) in the process of titration with the ammoniated cupric test, and thus interfere with an accurate determination being made. Under proper circumstances no such interfering colour is produced. About 2 grams was the quantity usually taken for analysis. This was boiled in a flask, with the use of the inverted condenser, with 50 c.c. of the potash solution, for half an hour, the vessel being agitated from time to time, so as to rinse down the particles attaching themselves to the glass above the liquid, and secure that none escaped solution. The contents of the flask were then poured into not less than 500 c.c. of methylated spirit, and the beaker was placed aside until the following day for the thorough settlement of the precipitate. The precipitate was now collected on a glass-wool filter plug, washed with alcohol, dissolved in hot water, and, after the addition of sulphuric acid to the extent of 2 per cent., boiled for an hour and a half with the use of the inverted condenser, or placed in the autoclave and submitted for half an hour to a temperature of 150° C. (about 300° F.). The acidified product was then neutralised with potash, made up to a known volume, thrown on to a dry filter, and finally titrated with the ammoniated cupric solution. The subjoined account gives a representation of the results obtained.

"*Egg Albumin.*—Prepared in some of the instances by precipitation with alcohol, and in the others by treating with water, faintly acidifying with acetic acid, and boiling.

"With the 10 per cent. solution of potash, the results showed the existence of a cupric oxide reducing power which, reckoned as glucose, averaged about 30 per 1000. With the 2 per cent. potash it stood at about 10 per 1000.

"*Vitellin from Yolk of Egg.*—About the same average figures were yielded as by egg albumin.

"*Proteids of Blood Serum.*—One observation, in which the figures stood at 16 per 1000 after the employment of 10 per cent. potash, and at 6 per 1000 after 2 per cent.

"*Proteid of Haricot Bean* (classed as a globulin).—Obtained by extracting with cold water, filtering the solution, and coagulating by faintly acidifying with acetic acid and boiling. The absence of starch was proved by iodine. 91 per 1000 constituted the figures obtained after treatment with 10 per cent., 77 per 1000 after 2 per cent., and 47 per 1000 after 1 per cent., potash.

"*Gluten from Wheat Flour.*—Washed till freedom from starch was shown by iodine. The figures given after the employment of 10 per cent. potash solution stood at 60 per 1000, and after 2 per cent. at 54 per 1000, in the case of one specimen; and in that of another, at 53 per 1000 after 10 per cent., 30 per 1000 after 2 per cent., and 24 per 1000 after 1 per cent.

"As I have said, I formerly looked upon the cupric oxide reducing product given by the process of analysis I have described as emanating from the presence of free glycogen. The evidence I have just adduced negatives this view. If free glycogen or starch had been the source of the reducing product, the treatment with potash would have produced no effect beyond dissolving the associated nitrogenous matter and placing it in a position to be separable by the agency of alcohol, and no difference would have resulted from varying the strength of the alkali or the length of time of contact. The conclusion, therefore, deducible is that the cupric oxide reducing product taking origin under the circumstances must be derived from some other source, and that the source lies in the cleavage or disruption of the proteid molecule itself.

"Besides the principles referred to above, from which, as I have shown, a cupric oxide reducing product is to be obtained by the cleavage action of potash, the subjoined have also, with the results specified, been subjected to treatment with potash of 10 per cent. strength.

"*Fibrin.*—Obtained by whipping freshly drawn blood, washing the stringy coagulum with water till colourless, immersing in alcohol for dehydration, and afterwards drying by exposure to the air. The cupric oxide reducing product yielded after treatment with 10 per cent. potash amounted in one case to 22·07 and in another to 22·72 per 1000, expressed as glucose.

"*Mucin.*—Obtained from the vitreous humour of sheep's and bullocks' eyes by precipitation with alcohol and subsequently drying. 27·4 to 29·6 per 1000 constituted the range of figures yielded.

"*Casein.*—Obtained from milk by heating and faintly acidifying with acetic acid. The coagulum after being washed was squeezed, dehydrated by alcohol, and treated with ether for removal of fat. The cupric oxide reducing product yielded amounted only to 2 to 4 per 1000. The difference here presented from the figures in every case previously given is very marked, and from this and other considerations the idea is suggested that casein may itself be a nitrogenous cleavage product, lactose constituting the complementary carbohydrate part.

"*Gelatin.*—It is to be noted as a point of difference from proteids that no cupric oxide reducing product is obtainable from gelatin. The gelatin sold under the designation of "French leaf" is the kind I have submitted to examination.

"In accord with the known effect of water at elevated temperatures in leading, to a greater or less extent, to the splitting up of glucosides, it is found that from proteid matter a certain amount of cupric oxide reducing product is similarly obtainable. After treatment of egg albumin with water at a temperature of about 150° C. (300° F.), I have obtained from the liquid a product which has given unmistakable evidence of possessing cupric oxide reducing power.

"*Preparation and Properties of the Cleavage Product, susceptible of Conversion into a Cupric Oxide Reducing Body (Sugar), derived from Proteid Matter.*

"I have shown how, through the quantitative analyses conducted with the ammoniated cupric solution, I was led to the discovery that a cupric oxide reducing product is obtainable from proteid matter. I have referred to my fruitless attempts to collect the cleavage product of the action of potash in quantity, and to the difficulty thereby created through the conflicting evidence presented. Subsequently an instance occurred in the course of the prosecution of my researches from which I learnt that the precise strength of the alcohol employed for precipitation after the boiling with potash constituted an item of greater importance in the process than I had previously realised. Looking, as I had originally done, on the precipitated material as glycogen, I had assumed, from the known sparing solubility of this body in alcohol, that as long as the existing strength of spirit was not under 60 per cent. full precipitation would be secured. It now became evident to me, however, that such was far from being the case, and that unless much stronger spirit were used only partial precipitation occurred.

"This information regarding the loss that may arise from the employment of a strength of spirit that I had previously regarded as sufficient, together with what I have shown to be the different extent of cleavage effect resulting from the employment of different strengths of potash, supplies the key to the explanation of the former want of success attending my endeavours to obtain the product in quantity. In operating upon large amounts of material, the circumstances are such as, without the knowledge of the requisite precautions to be observed, to be likely to lead to loss both from incomplete cleavage by the potash and incomplete precipitation by the spirit. Suffice it that I have now no difficulty in obtaining the product in any amount that may be required for the purpose of examination or experiment, and I will proceed to describe the steps of procedure I adopt with, for example, egg albumin, which has appeared to me to be the most suitable form of proteid to take as a representative of the group.

"The whites of twelve eggs, separated from the yolks, are broken up in an egg beater, or by whipping, and poured a little at a time into a large capsule of boiling water, acidulated with acetic acid to the point required for obtaining a satisfactory coagulation of the albumin. The water is strained off through muslin, and the coagulum washed, squeezed, and then placed in a flask with 20 grams of potash dissolved in a small quantity of water. The mixture is placed on the water-bath for two or three hours or set aside till the following day, with in either case an occasional shaking, to become liquefied. The measurement of the liquid is now taken, and potash added to the extent required to bring to a 10 per cent. strength. The flask having been fitted to an inverted condenser, its contents are boiled for half an hour. The liquid is next treated with acetic acid till rendered faintly acid, and after filtration concentrated on the water-bath to a bulk of about 100 c.c. in order to diminish the quantity of alcohol subsequently required. It is now in a somewhat viscid state, and if simply poured into spirit would sink as a coherent mass. To avoid this, a certain amount of spirit, insufficient for precipitation, is added, and aferwards the whole poured into a further quantity of about 2 litres to secure that the strength of alcohol is in excess of what is actually required. The material then separates out in a finely divided form, and on the following day will be found to have settled into a gummy mass at the bottom of the vessel, from which the spirit may be removed by simply pouring off. This constitutes the product from which a cupric oxide reducing body is obtainable, and which I formerly regarded as consisting of glycogen in a crude state.

"The properties of the material thus obtained are as follows :—

"In the dried state it forms a hard, glassy, resinoid mass.

"It is readily soluble in water, giving a clear solution.

"It yields no coloration with iodine.

"It possesses no cupric oxide reducing power.

"It is precipitable by alcohol, but alcohol of considerable strength is required for the purpose. In the presence of spirit of 85 to 90 per cent. it is in great part, if not completely, precipitated, settling down as an adherent, tenacious, gummy mass, from which the alcohol may be decanted off, and which may afterwards be worked up by stirring with a glass rod into a sticky material. With absolute alcohol, used freely, it is thrown down as a finely-divided white precipitate, without any tendency to coalesce into a gummy mass. When precipitated in the gummy form treatment with absolute alcohol exerts a dehydrating action upon it, causing it to harden and assume a crumbled, in place of a cohesive, state. In the presence of spirit of less than 85 per cent. strength, precipitation becomes more and more incomplete, and the precipitate produced by the weaker kind of spirit assumes a loose or non-adherent form. The incompleteness of precipitation by moderate strengths of alcohol is readily made apparent by the further precipitation that occurs upon the addition of more alcohol to the supernatant spirit.

"In its physical characters it presents a resemblance to Landwehr's 'animal gum.' Chemically it also resembles Landwehr's 'animal gum' in forming a copper compound on being treated with cupric sulphate and caustic potash, from which it is susceptible of recovery by the agency of hydrochloric acid and subsequent precipitation with alcohol. Landwehr's process is in substance as follows :—

"The cleavage product obtained from mucin by the action of dilute hydrochloric acid is dissolved in water, and to the solution are added a sufficiency of copper sulphate and excess of caustic potash. The precipitate containing the copper compound is separated by filtration, assiduously washed, dissolved in as small a quantity as possible of strong hydrochloric acid, and the solution poured into three times its volume of absolute alcohol. On placing the alcoholic liquid on the water-bath at about 60° C., a flocculent precipitate almost immediately begins to separate out, which constitutes the liberated "animal gum," and which, on being collected and subjected to appropriate treatment with sulphuric acid, gives origin to a cupric oxide reducing product.

"I have applied this process to the product under consideration derived from the action of potash upon albumin, and have found that it behaves throughout like Landwehr's 'animal gum,' and similarly yields a cupric oxide reducing body. Moreover, I have further found that this cupric oxide reducing body gives with phenylhydrazine needle crystals of glucosazone, of which I have obtained micro-photographs.

"The product, it may be finally remarked, possesses the property of diffusibility, a character in which it differs from the amylose carbohydrates—starch, glycogen, and dextrin. I have been unable to find any statement about the diffusibility or otherwise of 'animal gum,' but it is described as being a constant constituent of the urine, which may be regarded as indicative of its being of a diffusible nature.

"*Preparation and Properties of the Cupric Oxide Reducing Product (Sugar) derivable, through the preceding product, from Proteid Matter.*

"By the action of mineral acids, the first-formed product of which I have been speaking undergoes conversion into a cupric oxide reducing material. For effecting this conversion, I have for years past been in the habit of using 2 per cent. sulphuric acid. I was under the impression, as I have already said, that the product consisted of glycogen, and from observations upon the conversion of starch and glycogen into glucose, I had formed the opinion that sulphuric acid of the strength named best and most securely met the requirements. Boiling for an hour and a half with an inverted condenser was resorted to, unless the autoclave was used, in which case half an hour's exposure to a temperature of 150° C. was found to produce an equivalent effect, the results given by the two methods being practically identical. Where a quantitative analysis has constituted the object in view, the acid employed has been subsequently neutralised with potash. The sulphate of potash formed does not, under the circumstances,. occasion any inconvenience; but, should it be desired to collect the reducing product, the acid must be removed by precipitation, and this is best effected by the agency of barium carbonate. The filtrate from the barium sulphate and surplus barium carbonate being afterwards evaporated to dryness on the water-bath, the material in the state desired is yielded.

"With regard to the cupric oxide reducing product obtained, I had all along felt, from certain points connected with its manner of reducing the ammoniated cupric solution, that it did not consist of glucose, and I several times tried to get it carried higher in cupric oxide reducing power by the acid. I noticed with products which had been allowed to stand for some days or weeks after boiling with the acid that indications were afforded suggestive of the occurrence under examination of increased cupric oxide reducing action, but the evidence did not come out with sufficient distinctness to permit of my drawing any definite conclusion upon the point. Moreover, I had the fact before me that standing with potash was attended with an increased effect, and it seemed to me unintelligible that the same result should occur from standing with the acid. In reality, however, there is truth in what I observed, and I can now demonstrate that the body I obtained with the 2 per cent. acid is susceptible of being carried into one of much higher cupric oxide reducing power. What appeared to be anomalous is now cleared up. Through the potash, the capacity exists for an influence to be exerted upon the amount of cleavage product developed; and through the acid, the subsequent transformation may be influenced in such a manner as to give a product with an increased cupric oxide reducing power, and thus the semblance of an increase of material.

"It was from what I observed whilst working with the phenylhydrazine test that the suggestion was supplied which led to the next step of progress. Struck with the resemblance between the deposit (not crystalline, it is to be remarked) given by the reducing product derived from the action of the 2 per cent. acid and that given by lactose (the ordinary crystallised of commerce), I was led to renew my efforts in the direction of getting the body carried higher in cupric oxide reducing power. In experimenting with cellulose, I had found that but very slight action was produced by 2 per cent. acid, but that considerable effect followed the employment of acid of 10 per cent. strength, and it occurred to me as possible that the body I was dealing with might, as regards resistance to the action of acids, stand in a similar position. I was thus led to try the effect of 10 per cent. acid, and, as the result, found that the reducing power of the product became nearly doubled—in other words, became raised in the proportion of from between 50 and 60 to 100. Upon the strength of this result, I tried the effect of 50 per cent. acid allowed to remain in contact with the product for from one to three days, then diluted to 10 per cent., and boiled. The result stood about the same as after direct boiling with 10 per cent. acid. After 15 per cent. acid, also, a like result was obtained. As yet I have failed to carry the body to a higher stage of cupric oxide reducing power than that produced by 10 per cent. acid, but I am, nevertheless, not satisfied that the stage of glucose has been reached. Indeed, I am led to think that it has not.

"The effect of raising the cupric oxide reducing power is to give a semblance of a corresponding increase of material. The figures, therefore, representing the amount of cupric oxide reducing product expressed as glucose, obtainable from the various proteids as a result of the employment of 2 per cent. sulphuric acid, require, as a result of the employment of 10 per cent. acid, to be nearly doubled.

"The product obtained after the separation of the sulphuric acid by barium carbonate and evaporation to dryness possesses the following characters:—

"It presents the appearance of a sugary extractive, and possesses a pronounced baked-sugar odour.

"It is very soluble in water, only slightly soluble in absolute alcohol, considerably soluble in spirit of about 90 per cent. strength.

"It is readily diffusible.

"On being boiled with caustic potash (Moore's test), the solution darkens in colour.

"It dissolves hydrated oxide of copper in the presence of potash in excess (Trommer's test), without the production of a biuret reaction; at least, specimens are often procurable of which this can be said.

"With Fehling's solution it gives a strong and neat reaction, the reduced oxide of copper falling as a dense red precipitate.

Photo-engraving of micro-photograph of osazone crystals from the cleavage sugar from egg albumin. Magnified 400 diameters.

"Heated on the water-bath for from two to three hours with phenylhydrazine and acetic acid, it yields on cooling a crystalline osazone, presenting the form of needles aggregated into sheaves or brushes, or clustered in a radiating manner into dense round masses. Frequently the composite character of the round masses is only to be seen on close inspection of the circumference. Whilst a certain type of character is presented, variations within certain limits are noticeable, and sometimes an approach to a spike constitution is observable. The crystals are soluble in alcohol, from which they may be recrystallised.

"It gives with benzoyl chloride an insoluble compound, in accord with the behaviour of carbohydrates.

"With α-naphthol and excess of strong sulphuric acid it behaves like sugar in giving a deep violet colour, and leading on dilution to the formation of a violet-blue precipitate soluble in alcohol, ether, and caustic potash, with the production of yellow solutions, but insoluble in hydrochloric acid, a character by which, according to Molisch, the precipitate produced from sugar is distinguishable from that derived from peptone and various albuminous bodies.

"With thymol and excess of strong sulphuric acid, it, again, behaves like sugar, giving a deep red coloration, followed on dilution by the production of a carmine-red precipitate, soluble in alcohol, ether, and caustic potash, with the formation of pale yellow solutions, and in ammonia with the formation of a bright yellow solution. The precipitate, as in the case of the α-naphthol test, is found to possess the character of insolubility in hydrochloric acid.

"It is susceptible of being thrown down in combination with oxide of lead, and is afterwards recoverable from the compound. To demonstrate this, the aqueous solution is first treated with neutral acetate of lead and filtered. Ammonia is then added to the filtrate, and afterwards lead acetate and ammonia as long as further precipitation occurs. The precipitate, which contains the sugar compound, is collected and washed with water until by test-paper it is shown to be free from ammonia. The sugar is now liberated by dissolving the precipitate in acetic acid, and the lead got rid of by precipitation with sulphuric acid. From the sugar thus recovered the usual reaction is obtainable with Fehling's solution (provided the ammonia has been fully removed); and also the crystalline osazone, with phenylhydrazine. Subjoined is a micro-photographic representation of the crystals I have obtained. On comparing them with those previously represented, it will be seen that a somewhat modified form is presented as the result of the process that the sugar has passed through.

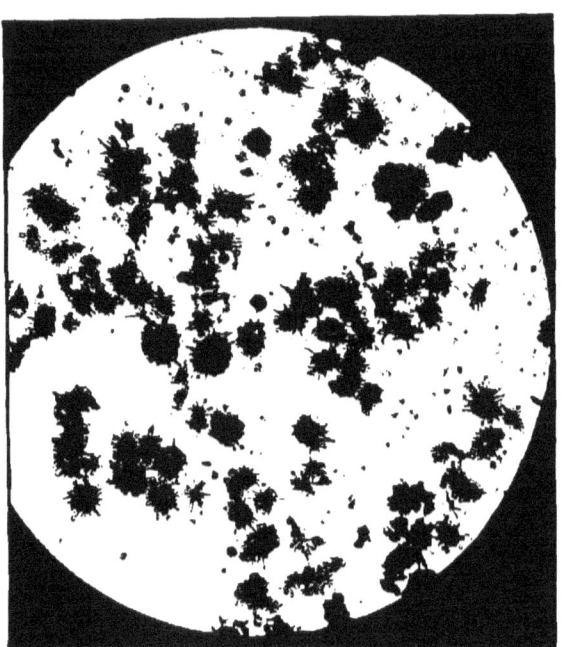

Osazone crystals from the cleavage sugar from egg albumin recovered from the lead compound. Magnified 400 diameters.

"From this assemblage of positive characters, it appears to me that there can be no doubt that the cupric oxide reducing body obtainable from proteid matter consists of sugar. It is to be stated, however, that I have not yet obtained it in a form to prove fermentable, and Dr. Sheridan Lea, who has kindly undertaken its examination with the polarimeter, informs me that he has failed to notice any rotation of which he could speak with certainty. If anything, there was, he adds, a tendency to lævorotation, but amounting to not more than $0°·1$.

"As regards both fermentability and optical activity, it is known that sugars exist without the possession of these properties. Landwehr's 'gummose,' which stands closely related to my cupric oxide reducing product, if it is not actually identical with it, constitutes an example of a sugar that does not ferment. With regard to optical activity, not only are sugars known which do not exhibit this property,* but there is the further point of consideration that a rotatory power which may exist may be neutralised, and thus masked by a rotation in the opposite direction due to the incidental presence of other bodies.

"In addition to the evidence derivable from the positive characters that have been mentioned as possessed by the cupric oxide reducing product, there is the corroborative evidence to be taken into account supplied by the characters of agreement observable between my primary non-reducing product and Landwehr's 'animal gum.' As I have already shown, the product in question can be thrown down, precisely like 'animal gum,' in combination with copper oxide, and is afterwards recoverable and convertible into a cupric oxide reducing body, which, as I have further said, yields with phenylhydrazine a crystalline osazone. Two kinds of crystals are perceptible in the subjoined micro-photograph (p. 42), one consisting distinctly of the long needles of glucosazone. In its ready mode of separating out, the osazone, it may also be said, was observed to agree with that derived from glucose. A 10 per cent. strength of sulphuric acid was employed in obtaining the cupric oxide reducing product operated upon.

"In closing this communication, I feel it due to my assistants, Mr. Rowntree and Mr. Sian, to acknowledge the zeal with which they have

* It is known, as a practical detail, to those who are engaged in the sugar-cane industry that the glucose present in the living and growing cane is not possessed of optical properties, or, at least, if it does possess optical properties, that they must be of a nature to exactly neutralise each other, for if the freshly-cut cane is at once put through the mill the glucose contained in the extract is found not to interfere with the polarimetric estimation of the cane sugar. With the glucose, on the other hand, recognizable after the cane has been cut or injured, the circumstances stand otherwise. The juice here contains glucose of an optically active nature, consisting of ordinary invert, or transformed cane, sugar. This may greatly exceed in amount the inactive glucose present in the fresh cane; and, in proportion to its amount the estimation of the cane sugar is interfered with.

Upon the question of optical activity Emil Fischer remarks that experience, in connexion with the bodies belonging to the sugar group, confirms the view that for every optically active substance there exists an isomeric form of opposite optical activity, and that the two compounds combine together to form an inactive modification.

helped me in the work that has been performed. I further consider it due to state that the facilities afforded by the Research Labora-

Osazone crystals from the sugar obtained from the primary non-reducing product after recovery from its copper compound. Magnified 400 diameters.

tories of the Colleges of Physicians and Surgeons have largely contributed to enable me to carry on my investigation work to the point that has been attained."

The above transcript represents the extent to which my knowledge had reached when I communicated my results to the Royal Society in May last. I thought it probable that by further work, through the aid of the copper precipitation of my primary non-reducing product, sugar might be found to be obtainable in a sufficiently pure form for submission to combustion analysis. I have exerted my endeavours towards the achievement of this object. The successful precipitation appears to rest upon the employment of delicately balanced quantities of the agents used; but, as stated in the communication above set forth, I have obtained from the copper precipitate a product which on being subjected to the inverting influence of sulphuric acid has acquired cupric oxide reducing power, and has given with phenyl-hydrazine the osazone crystals represented by the photo-engraving inserted. Repetitions of the process, however, have not been always attended with success; and there is this important consideration, that, whatever may be the case with Landwehr's animal gum obtained from mucin, my product, obtained by the action of potash upon albumin after subjection to the copper precipitation process, is accompanied with other material which comes out in the spirit on warming in a flocculent form, precisely as is done by animal gum. This establishes a difficulty which entirely frustrates the attainment of the object to which I thought the copper precipitate might prove applicable.

Cleavage of Carbohydrate from Proteid Matter by the direct action of Sulphuric Acid.

An extension of research upon modified lines has led to another important step being effected. Through the advance made, the preliminary production of the non-cupric oxide reducing material by the agency of potash is no longer needed. It turns out, in fact, that the cleavage of the proteid molecule, with the liberation of sugar, can be brought about by the direct treatment of albumin with sulphuric acid. Thus the process for demonstrating the glucoside constitution of proteid matter becomes very much simplified and shortened.

I tried, some time ago, the effect of the direct action of sulphuric acid upon the proteid matter of muscular tissue. I had not then worked with phenyl-hydrazine, and I could make out nothing definite through the employment of the copper test, on account of the obscuring effect of the peptone reaction, although I felt strongly impressed that evidence was afforded of the occurrence of a reducing action. I made attempts to separate the peptone, but failed to find a satisfactory means of doing so. Later, whilst working at the osazones, it occurred to me to apply this method of sugar recognition to the product of the direct action of sulphuric acid on proteid matter. The result obtained was a copious production of well shaped osazone crystals.

For the application of the process, 10 grams of purified, dried, and finely divided (passed through a 90 to the linear inch sieve) egg albumin constitutes a convenient quantity to take. As sugar is originally present, which would remain undestroyed by the action of the acid, it is necessary that care should be exercised to secure complete purification. This, it may be considered, is effected by the plan adopted, consisting of precipitating by and boiling with alcohol, subsequently extracting with boiling water, and finally with boiling alcohol.

The material to be dealt with is placed in 50 c.c. of 10 per cent. sulphuric acid and exposed to heat in the requisite manner. The form of osazone crystal obtained, as will be presently shown, varies with the extent of influence exerted by the degree and duration of the heat brought to bear. The further the action of the acid is pushed, the quicker the osazone separates out and the more distinctly acicular, in brush-like or stellate aggregations, the character of the crystals formed. The less complete the action, the more ball-like the form of crystal presented and the less speedy the deposition. It appears to me, from all I have seen, that the different forms of crystals arise from corresponding modified forms of sugar; and, that in the first manifestation of the power of crystallising the character assumed is that of a nebulous or moss-like radiate body. In the case of the sugars from other sources, it is found that, the lower the cupric oxide reducing power, the more, apparently, does the osazone crystal deviate from the acicular form belonging to the osazone derived from glucose.

The suggestion arising out of these considerations is that probably modifications of sugar exist beyond those which are now definitely recognised and described, and that the initial product cleaved off from proteid matter by the action of sulphuric acid is more or less widely removed from glucose, but becomes progressively carried towards it by prolongation of the action.

Doubtless much remains to be learnt regarding the extent to which varying molecular combinations may occur to produce different varieties of sugar. Just as cane sugar may be conceived to be constituted of a linking together of a dextrose group and a lævulose group (with concurrent dehydration); maltose, of a linking of two dextrose groups; milk sugar, of a dextrose group and a galactose group; and raffinose, of three different glucose groups—dextrose, lævulose, and galactose: so there possibly may be other linkings of sugar groups of different degrees of hydration and in varying multiple proportions, giving rise to a large extension of the varieties of sugar beyond those at the present time particularised by chemists.

A foundation for this hypothesis is presented by the view entertained by authorities of eminence with regard to what occurs in the successive steps of hydrolysis attending the transformation of starch into maltose. Amylodextrin, for instance, is spoken of as a body consisting of one maltose ($C_{12}H_{22}O_{11}$) group linked with six amylose ($C_{12}H_{20}O_{10}$) groups, and maltodextrin as a body consisting of one maltose group linked with two amylose groups.

After the requisite exposure to heat has been completed, the sulphuric acid employed is nearly neutralised by baryta, and neutralisation finished off with barium carbonate. The liquid is then brought to the bulk required, and, after the addition of phenyl-hydrazine and acetic acid, is heated on the water-bath for at least an hour. Crystals

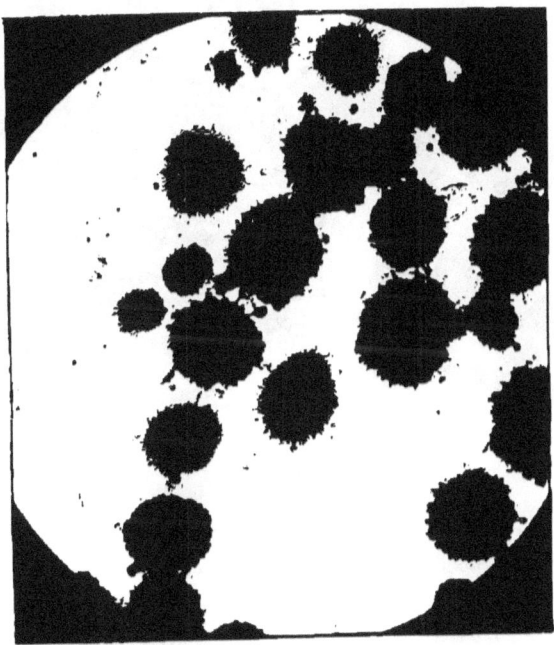

Osazone crystals from cleavage sugar obtained from egg albumin by direct action of sulphuric acid, after limited extent of action. Magnified 400 diameters.

are found to make their appearance, sometimes whilst on the water-bath, but more frequently not till cooling commences, and to go on separating out for some hours afterwards. If the hot solution is poured into a cool glass the crystals fall in quite a shower-like manner.

Osazone crystals from cleavage sugar obtained from egg albumin by direct action of sulphuric acid, after more prolonged action. Magnified 400 diameters.

From the osazone obtained the sugar is susceptible of recovery by following the procedure described by Fischer. The osazone crystals are collected on a glass-wool filter, washed with water, and dried. They are then treated with a little strong hydrochloric acid (sp. gr. 1·19), quickly warmed to 40° C., at which temperature the mixture is kept for one minute, next cooled to 25° C., and then allowed to stand for ten minutes. The phenyl-hydrazine hydrochloride which has been formed is now separated by filtration through glass wool, and is washed on the filter with a little concentrated hydrochloric acid. The filtrate contains a body corresponding with the glucosone described by Fischer. For the separation and purification of this body, the acid is first neutralised with lead carbonate, and, after filtration, the liquid treated with animal charcoal. Excess of baryta-water is next added, the glucosone-like body being thereby thrown down in combination with lead hydrate, from which it is subsequently separated by the employment of dilute sulphuric acid. After the excess of sulphuric acid has been removed with barium carbonate, the liquid is concentrated to a syrup by evaporation at a low temperature. The body thus separated has the capacity, like sugar, of uniting with phenyl-hydrazine, and the following is a representation of osazone crystals obtained from it.

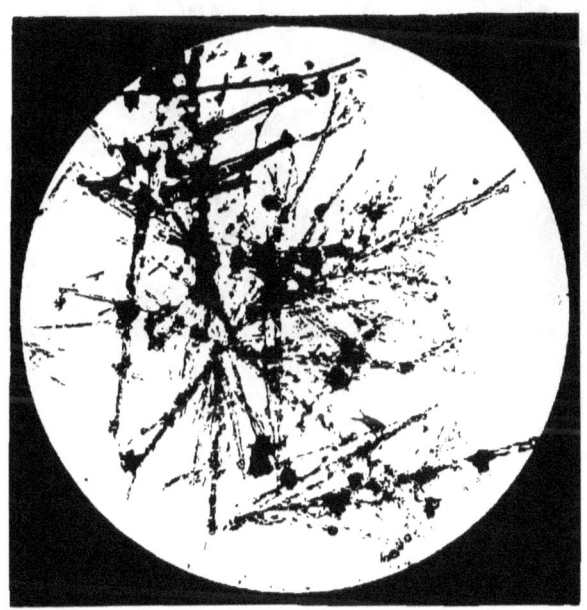

Osazone crystals from the intermediate body obtained in the process of recovery of sugar from the cleavage-sugar osazone. Magnified 400 diameters.

From the intermediate body sugar is obtained by reduction with zinc-dust and acetic acid. The recovered sugar is found, after removal of the zinc present, to give a clean and definite reaction with the copper test, attended with a distinct deposition of red oxide particles. It also yields an osazone with phenyl-hydrazine. The following is a photo-engraving from a micro-photograph of crystals so obtained.

Osazone crystals from sugar recovered from cleavage-sugar osazone. Magnified 400 diameters.

The melting point of an osazone gives information which renders assistance in the identification of its sugar. I have conducted several observations, with the view of determining the melting point of the osazone from the cleavage sugar from proteid matter, and find that it may be stated to stand at about 189—190° C. It thus more closely approaches the melting point of galactosazone (190—193° C.) than that of glucosazone (205° C.).

From what has preceded, it will be perceived that the phenyl-hydrazine test has proved of immense value in the pursuit of the investigation which has revealed the glucoside constitution of proteid matter. It has not only served to corroborate the indications of the copper-reduction test, but has proved susceptible of application where this test, from the presence of interfering matter, has not been satisfactorily available. For instance, the product from the direct action of sulphuric acid upon albumin, when treated with the copper test, gives no reaction from which, taken alone, any reliable information could be drawn. From the biuret reaction produced by the peptonoid matter present a masking influence is exerted, and the ammonia generated interferes with the deposition of sub-oxide. With the knowledge supplied through the phenyl-hydrazine reaction it is not difficult now to recognise, notwithstanding the obscuring effect of the biuret reaction, that a reduction of the copper test in reality occurs. What is observed is that the blue colour imparted by the test liquid is, on boiling, removed, without the formation of any precipitate. That this disappearance of colour is actually due to reduction is shown by the return of colour which occurs from re-oxidation, when the contents of the tube are well shaken in contact with air and allowed to stand; and, further, by the fact that, under prolonged boiling, the ammonia produced may be so far dissipated that the deposition of a certain amount of reduced oxide is permitted to take place. Had it not been for the masking action of peptonoid matter, to which I have been referring, I venture to think that the liberation of sugar from proteid matter, looking at the simple and easy manner in which it is brought about by the agency of sulphuric acid, would long ere now have stood in the position of an established fact.

Cleavage of Carbohydrate from Proteid Matter by Digestive Ferment Action.

Another means can be shown to exist whereby the cleavage of carbohydrate from proteid matter can be effected. By alkalis it is cleaved off in a form devoid of cupric oxide reducing power, and likewise of the property of producing an osazone on treatment with phenyl-hydrazine, but susceptible of acquiring these properties through the converting influence of sulphuric acid. By sulphuric acid, applied direct, the carbohydrate is liberated in the form of a reducing sugar. By the agency of proteolytic ferment action the same effect is produced as by sulphuric acid, and we have here a point of much physiological importance presented. Purified egg albumin yields, under pepsin digestion, a product which gives characteristic osazone crystals with phenyl-hydrazine. In conducting the experimental work upon which this statement is based, I found that precautions require to be taken to secure that no fallacy is allowed to creep in. In the first place, the sugar of the egg albumin should be thoroughly removed by successive extractions with boiling water, or—what is better—by precipitating, in the first instance, with boiling alcohol, and, afterwards, successively extracting with boiling water and alcohol; and, in the next place, it must be ascertained that the pepsin employed does not contain impurity, as many samples of pepsin do, to constitute a source of osazone production.

Now, the result of experimenting with egg-albumin, which has been thoroughly deprived of free sugar, and pepsin, which, taken alone and exposed to contact with the acid as a counterpart experiment, gives a negative result, is that osazone crystals are obtained. Before concluding, however, that these crystals are derived from cleavage effected actually by digestion, another counterpart experiment requires to be performed, consisting in the exposure of the egg albumin to contact with the acid and water without the addition of the pepsin. Treatment, it is found, of the product of this procedure with phenyl-hydrazine yields no crystals of osazone, thus proving that the acid alone is not the operative agent. Indeed, with a strength of acid of only 0·2 per cent. and exposure to the moderate temperature required for digestion, no effect from the acid alone could be reasonably looked for.

From the evidence presented it is thus permissible to conclude that in the process of digestion the carbohydrate constituent of the proteid molecule is set free, and it will be noticed that it presents itself to our view under the form of sugar. Subjoined is a photo-engraving of a micro-photograph of osazone crystals given by the cleavage sugar arising from the digestion of egg albumin. Digestion was allowed to proceed for $8\frac{1}{2}$ hours at 54° C., the albumin in the coagulated, dried, and pulverised state in which it was used proving very resistant to digestive solution.

GLUCOSIDE CONSTITUTION OF PROTEID MATTER. 51

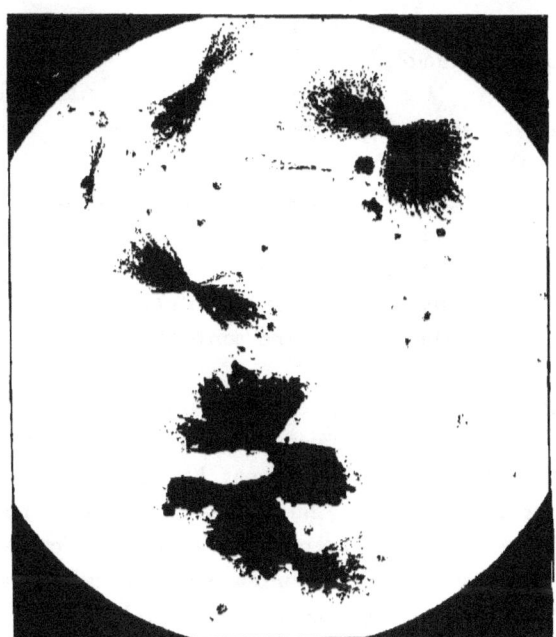

Osazone crystals from the cleavage sugar given by digesting 10 grams egg albumin with 0·4 gram pepsin and 100 c.c. of 0·2 per cent. hydrochloric acid. Magnified 400 diameters.

Conclusion from Array of Evidence presented.

From what has preceded it is seen that a cleavage product is in different ways obtainable from proteid matter which possesses characters presenting the closest points of identity with those of sugar. The reaction, especially, which it gives with phenylhydrazine, embracing not only the crystalline form and mode of separating out of the osazone obtained, but also the manner in which it subsequently behaves, affords the strongest presumptive evidence that sugar is present. There can be no doubt about the fact that an osazone is produced, and, with the knowledge that we possess regarding the derivation of osazones, it is not, it may be considered, permissible, under the whole circumstances of the case, to look to anything else than sugar as the source of the osazone given by the cleavage product from proteid matter.

The negative evidence, based upon optical inactivity and nonfermentability, cannot be considered to invalidate the positive evidence that is adducible.

Optical activity is a property in connexion with which much diversity exists, some sugars being dextro-rotatory, others lævo-rotatory, others inactive, from the neutralising effect of the presence of varieties with opposite rotatory powers, and, again, others absolutely inactive, and not to be resolved into component active sugars. Further, a sugar with rotatory power in one direction is susceptible of being converted into a sugar with a rotatory power in the opposite direction, as is exemplified by the convertibility of dextrose into lævulose (Fischer).

Fermentability is a property similarly circumstanced with regard to diversity. Sugars, it is known, exist, even amongst the glucoses, which are not fermentable, and it may happen that one form of a particular group having one kind of optical activity is fermentable, whilst another form, with an opposite kind of optical activity, is not. For instance, ordinary lævulose, which is lævo-rotatory, ferments, whilst a dextro-rotatory form of lævulose which has been artificially obtained is not fermentable. A similar difference is observable in the dextrose group, the ordinary or dextro-rotatory form of dextrose being fermentable, whilst the lævo-rotatory form is not (Fischer). Whatever may subsequently prove to be attainable, it is not contended that the sugar from proteid matter has yet been brought into the form of glucose—indeed, the evidence is to the effect that it has not. If, therefore, so resistant to passage into glucose, there is nothing surprising in its being resistant to the fermenting influence of yeast.

Whilst the recognition of the glucoside constitution of proteid matter was entirely founded upon the information drawn from its subjection to disintegration, the information derivable from the opposite consideration fits in and supplies confirmatory testimony. It will, in what immediately follows, be shown, upon irrefutable grounds, that in the construction of proteid matter by the synthetic power of protoplasmic chemistry, carbohydrate matter is a participating agent. If carbohydrate matter thus enters into the construction of the proteid molecule, it only stands in harmony with what might be expected, that it should be susceptible of being again liberated with the occurrence of disintegration.

Synthetic Formation of Proteids by the Incorporation of Carbohydrate with other Matter.

I will pass now to the consideration of the mode of origin of proteid matter, and as we proceed it will be found that what has to be stated gives support to the view concerning its constitution that I have enunciated upon the strength of results yielded by analytical investigation. It will be of advantage, as a preliminary step, to see what light is thrown upon the matter by the artificial production of glucosides.

Various glucosides have been artificially produced. Berthelot found that when glucose is heated for a number of hours at 100—120° C. in contact with acetic, butyric, and other organic acids, the two enter into direct combination, with an elimination of the elements of water. From the synthesised compounds thus obtained the carbohydrate may be again split off by heating with an alkali, and it is interesting to note that it reappears, not as glucose ($C_6H_{12}O_6$), but in a state of diminished hydration, viz., as glucosan ($C_6H_{10}O_5$). In fact, an illustration is here afforded, not only of the production of a glucoside by the incorporation of a carbohydrate with another body, but also of the subsequent separation of the carbohydrate in a modified form.

If we look now to the production of proteid matter by the synthesising action of living protoplasm, we meet with evidence unmistakably showing that the production may arise from the combination of carbohydrate with elements derived from nitrogen- and sulphur-containing compounds.

The facts connected with the growth of the yeast cell may be adduced in substantiation of what has been stated. Life is here carried on in so simple a manner that it is not difficult to express in precise terms the phenomena that occur, and the conditions as regards materials supplied for growth may be so arranged as to furnish a demonstration of the application of carbohydrate to the construction of proteid.

As Pasteur showed, many years back, yeast cells freely multiply in a medium containing sugar, ammonium tartrate, and the ash of yeast. The growth that here takes place implies a growth of cell protoplasm, and with it a corresponding formation of proteid matter, the only source for which is the group of principles contained in the surrounding medium. Although it may be reasonably inferred that the carbon, and the hydrogen and oxygen, of the sugar contribute to the formation of the proteid, the proof of such being the case is not absolute, seeing that all the elements in question are contained in the tartaric acid of the ammonium tartrate. It is possible, however, to supply a medium in which the growth will similarly take place, and which contains no carbon compound except the sugar. Such a medium is mentioned by Sachs* as having been devised by Adolf Mayer, and is stated to consist of water, sugar, ammonium nitrate (in place of the ammonium tartrate of Pasteur's liquid), acid potassium phosphate, tribasic calcium phosphate, and magnesium sulphate. When yeast is grown in this medium, it is self-evident that the whole of the carbon required to form proteid must be derived from the sugar. The example thus affords incontrovertible testimony that carbohydrate, in the presence of the other requisite matter, contributes, under the influence of the chemical power with which living protoplasm is endowed, to the construction of proteid.

The nitrogen supply needed for co-operating with the carbohydrate in the production of proteid may, as we have seen, be derived, in the case of the growth of a simple organism like the yeast cell, from even an inorganic source. In the higher vegetable organisms, where nutrient substances have to be transported to distant seats of utilisation, the nitrogen is supplied in a more elaborated form—very generally, it would appear, in the form of what is called asparagin, a crystallisable, diffusible principle, allied as an amido-compound in constitution to leucin and tyrosin, and, like these principles, susceptible of taking origin from the splitting up of proteid matter. Asparagin is found widely diffused through the vegetable kingdom. It is recognised by vegetable physiologists as a factor extensively concerned in the formation of proteid by combination with carbohydrate matter and the sulphur of a sulphur-containing body. It is especially to be found in the growing parts of plants, and appears to be formed from stored nitrogenous material as a principle adapted for transportation from the seats of reserve to those of active meta-

---

* 'Lectures on the Physiology of Plants,' by Julius von Sachs, translated by H. Marshall Ward, p. 383: Clarendon Press, Oxford, 1887.

bolic change, where a re-combination with carbohyrdate matter is asserted to take place.

The view represented, implying the joint participation of asparagin and carbohydrate in proteid formation, is founded upon observations which show that when the conditions are such as to permit of carbohydrates being present in abundance no asparagin is to be found at the seat of growth, whilst, if means are taken to limit their supply, as, for instance, by experimenting with slips grown in the dark or with the exclusion of carbonic acid, asparagin is to be readily detected. It can subsequently be made to disappear by exposing the shoots to strong light, and the disappearance is coincident with the formation of carbohydrates on the one hand, and the production of proteid matter on the other. The natural conclusion to be arrived at is that the asparagin combines with the carbohydrate and becomes used up in contributing to the formation of proteid matter.

Proteid Matter and the Deposition of Carbohydrate in the Vegetable Organism.

I have spoken of the participation of carbohydrate in the construction of proteid matter. I will now pass to the consideration of the converse aspect of the question and speak of its splitting off from proteid matter as an occurrence connected with the operations of life. It is noticeable that starch, cellulose, glycogen, &c., are deposited in the interior of the living cells of growing parts without having pre-existed as such. There is no doubt, it may be considered, that they are derived from sugar conveyed to the part in the juices of the plant. It is possible, it must be admitted, that they may take origin by direct transmutation of this sugar through an influence exerted upon it by living protoplasm, but it has been suggested as more probable that their production is the result of changes in the protoplasm itself. Sachs, after contending that the processes of primary starch formation by the chlorophyll corpuscles and secondary starch deposition by the protoplasmic starch-forming corpuscles from pre-existing sugar are carried out in essentially the same way, writes as follows[*] with reference to the former:—"It is, indeed, not impossible that certain more immediate constituents of the green plasma itself take part in

[*] *Loc. cit.*, pp. 317—318.

the process—that decompositions and substitutions, for example, take place in the molecules of the green protoplasm. This possibility obtains some probability from the observation that in many (not all) cases the chlorophyll substance gradually decreases in quantity and at length disappears entirely, while the starch-grains are growing in it." . . . "I hold it as probable that in this process the proteid substance of the assimilating chlorophyll itself co-operates and undergoes a change."

The purport of what is here said is that proteid plays a participating part in the process of deposition of carbohydrate as fabric and as store material—cellulose, starch, &c. It is true, the grounds of reasoning, as yet within our reach, do not supply the tangible evidence of the splitting off of carbohydrate from proteid matter by disruptive metabolism that is supplied of its entry by the exercise of constructive metabolism. Viewing all the circumstances, however, it is pardonable to surmise that the one stands as a natural corollary to the other. Carbohydrate is a co-operating factor in the construction of proteid. It is deposited in a form different from that in which it reaches the seat of deposition—a form, even, which may be said to present a semblance of organised structure, and it is undoubtedly through the agency, and only through the agency, of proteid as living protoplasm that the deposition is effected.

With all this, the discovery, arrived at quite independently by analytical manipulation, of the glucoside constitution of proteid matter agrees, and even might naturally be expected to follow. It supplies just the link that is wanted to fit in with the other links and complete the chain.

Under the view set forth, proteid matter, through its glucoside constitution, becomes of functional import in a manner, and to an extent, not hitherto definitely thought of. That its carbohydrate should be susceptible of cleaving off under exposure to certain influences (it may be considered that ferment agency constitutes one of them) is only in accord with the condition that is known to exist in the case of other glucosides.

The cellulose, starch, &c., deposited as above represented, are in a lower state of hydration than the sugar reaching the seat of the deposition. The position may or may not be analogous, but in any case it is an interesting circumstance that the carbohydrate derivable from artificial glucosides is, as I have previously stated (p. 53), similarly of a less hydrated form than the sugar employed in their construction.

These considerations, drawn from vegetable physiology, will be found later on to render service in studying the relations of carbohydrate matter in the animal system.

DESCRIPTION OF ANALYTICAL STEPS OF PROCEDURE.

A DESCRIPTION requires now to be given of the methods which experience shows to be necessary for effecting the recognition and estimation of the different carbohydrates which may be present in animal products. It is obvious that these products are, without preparation, not in a fit state for examination, and before any tests can be applied, it is necessary to obtain suitable liquids for the purpose. This preliminary treatment must be conducted in such a manner as to be unattended with a destruction of carbohydrate, or with a change in the nature of the carbohydrate beyond such intentional change as conversion into glucose.

Where it is simply required to detect the presence of sugar, without reference to its nature and amount, a process of aqueous extraction may be resorted to for obtaining a liquid for testing. Thus, for instance, by boiling the product with a sufficient quantity of sodium sulphate crystals or by boiling after exact neutralisation with acetic acid, the colouring and albuminous matters are completely separated, and a perfectly clear and colourless liquid containing the sugar is obtained.

This method is an excellent one for purposes of qualitative testing, but it is inappropriate for use where it is required to ascertain the nature and (unless the nature is previously known) the amount of the sugar existing. For obtaining this information it is necessary to determine the cupric oxide reducing power of the product in its initial state, and again after subjection to the glucose-forming influence of sulphuric acid. With an aqueous extraction, figures due to the presence of carbohydrate matter other than sugar would be given after treatment with sulphuric acid, and error would thus be introduced into the result. On this account an agent like alcohol, which takes up sugar but not the other carbohydrate matter, must, as will be presently more particularly mentioned, be employed for the purpose of extraction. If the material present is known to consist of glucose or any other particular form of sugar, a simple determination of its cupric oxide reducing power, without the employment of acid, will suffice for ascertaining its amount; and the sodium sulphate method, as affording an easier and more speedy process, may then be made use of for obtaining a liquid for quantitative examination. The following is a description of the procedure in this case to be adopted, where blood, taken as an example, is the product dealt with.

About 40 grams of sulphate of soda in small crystals are weighed out in a beaker or capsule. About 20 c.c. of the blood intended for analysis are then poured upon the crystals, and the beaker and its contents again weighed. In this manner the weight of the blood taken is ascertained. The blood and crystals are well stirred together with a glass rod, and about 30 c.c. of a hot concentrated solution of sulphate of soda added. The vessel is placed over a flame guarded by wire gauze, and the contents heated till a thoroughly formed coagulum is seen to be suspended in a clear, colourless liquid, to attain which actual boiling for a short time is required. The liquid is next to be separated from the coagulum. This is done by straining through a piece of muslin resting in a funnel. Some of the hot concentrated solution of sodium sulphate is then poured on the coagulum, well stirred up with it, and the whole thrown on the piece of muslin. The coagulum is squeezed, and, to secure that no sugar is left behind, is returned to the beaker, and the process of washing and squeezing repeated.

The somewhat turbid liquid is next thoroughly boiled, and then filtered through filter paper, when a perfectly clear filtrate is obtained. The washing of the beaker and filter with the hot concentrated solution of sodium sulphate completes the process of preparation. The liquid yielded is in an excellent condition for the quantitative determination of the sugar, by either of the methods, gravimetric or volumetric, presently to be described.

The process, as I have already stated, is only suitable for application where the nature of the sugar present is known. Where its nature is not known, a single determination, on account of the different degrees of cupric oxide reducing power possessed by the various sugars, suffices only to reveal the amount of reducing power belonging to the product, there being nothing to show whether this is representative of a certain amount of glucose or a larger quantity of some body or bodies of lower cupric oxide reducing power.

To acquire information regarding the nature of the material, a method of extraction must be resorted to which will admit of the sugar being obtained free from admixture with other carbohydrate matter (glycogen, &c.). By means of alcohol the desired separation is effected, the sugar being taken up in solution, whilst the other carbohydrate matter remains in the coagulated residue. The extracted carbohydrate being subsequently subjected to the converting influence of sulphuric acid, the determination of its cupric oxide reducing power, both before and after this treatment, affords an insight into the nature of the sugar present. If the two results coincide, the existence of glucose is indicated. If the figures yielded by the first determination are lower than those given by the second, it is indicative of the existence of a body or mixture of bodies of lower cupric oxide reducing power than glucose.

A single determination of cupric oxide reducing power is thus of no actual value, unless the particular form of sugar present is known. A failure to grasp the importance of this fact is calculated to lead to more or less extensive error in estimating quantities of carbohydrates. Maltose, for instance, has a reducing power of only 61 as compared with that of glucose taken at 100. Representative solutions, therefore, containing respectively 61 parts by weight of glucose and 100 parts of maltose, would exert precisely the same cupric oxide reducing effect, and would appear, as the result of a single determination, to contain equal amounts of carbohydrate. If each, however, were then boiled with dilute sulphuric acid and again examined, the results obtained would be widely different, that from the glucose-containing liquid remaining as before, whilst the figures yielded by the other would show a proportional advance of from 61 to 100. In the case of a low reducing mixture of maltose and dextrin, such as results in the early stages of the transformation of starch and glycogen, the error arising from reliance on a single determination of the cupric oxide reducing action would be considerably greater, and would, therefore, lead to the carbohydrate being estimated at a mere fraction of the amount actually present.

The details of the process of sugar extraction by means of alcohol, and the subsequent treatment of the residue for the collection and estimation of the carbohydrate, other than sugar, obtainable in the examination, for instance, of blood, which was taken as an example before, are as follows:—

About 30 grams of defibrinated blood are poured into about 400 c.c. of alcohol. Care must be taken that the spirit employed is free from impregnation with anything capable of subsequently exerting a cupric oxide reducing action. Ordinary methylated spirit may be used if it has been previously redistilled. If any doubt is felt as to its character, a control experiment may be performed with a portion of spirit to which blood has not been added.

The sugar present in the blood passes into solution in the alcohol. It is held, however, very tenaciously by the coagulated matter, and requires to be carefully extracted from it by repeated washing and pressing. The blood and spirit are well stirred together, and it appears to be of advantage towards obtaining as slightly coloured a product as possible for subsequent testing to allow the mixture to stand until the following day. After being exposed for a time to the heat of a water-bath, for the purpose of promoting the agglomeration of the precipitate, the alcohol is strained off through muslin which has been previously deprived by thorough washing with boiling water of its dressing of starchy matter. Washing with alcohol is performed, and the coagulum in the muslin is subjected to forcible squeezing in a suitable press. The residue, which by this process is converted into a dry cake, is pulverised in a mortar, mixed with fresh spirit, boiled over the water-bath, and again strained and pressed. The process is repeated once more. After this third extraction, the residue may be considered to be free from sugar, an aqueous extract failing to exert any cupric oxide reducing action.

Two extractions with alcohol might prove sufficient, if carefully made, but it is safer to use three. After the first extraction, there is evidence that a considerable quantity of sugar still clings to the coagulum. This is rendered apparent by the following results of actual observation. Upon one occasion 50 c.c. of sheep's blood were extracted with alcohol: the first extract contained 0·015 gram of glucose, the second 0·004 gram, and the third no definite amount. Again, some sugar from diabetic urine was added to sheep's blood, and 50 c.c. were taken: the first alcoholic extract was found to contain 0·144 gram of glucose, the second 0·009 gram, and the third nothing definite. These examples suffice to show the necessity of attention being given to secure complete extraction of the sugar.

The three alcoholic extracts, incorporated together, are now acidified with acetic acid, then heated to near the boiling point over the water-bath, and finally filtered through ordinary filter-paper. Whilst the residue on the filter is reserved to be subsequently dealt with in conjunction with the bulk of residue previously collected, the filtrate is evaporated nearly to dryness on the water-bath, and then briskly boiled with a few crystals of sulphate of soda or a few c.c. of its saturated solution, for the purpose of causing the colouring and fatty matters to agglomerate so as to be easily removed by filtration. The liquid is filtered off, and, to avoid the inconvenience of the crystallisation that may take place when a filter-paper is used, it is found advantageous to have recourse to filtration through a plug of glass-wool. The beaker and filter are then washed with the hot solution of sulphate of soda. Attention, it may be stated, should be given to secure that the sulphate of soda employed is free from any admixture of carbonate, as this impurity may vitiate the result by exerting, through its alkalinity, a destructive influence upon the sugar, besides leading to the production of a more coloured liquid for testing.

The filtrate obtained presents the product of alcoholic extraction in a suitable form for the quantitative determination of the sugar. It is made up with water to a known volume—conveniently to 100 c.c.— and then divided into two equal portions. One portion is at once examined, and its cupric oxide reducing power ascertained. The other is submitted, after the addition of sulphuric acid to the extent of 2 per cent., to appropriate treatment for the conversion of the carbohydrate matter present into glucose. Another determination of cupric oxide reducing power is subsequently conducted, in order that a comparison may be made between the two. As a preliminary step, however, the neutralisation of the sulphuric acid that has been used requires to be effected. For this purpose a strong solution of caustic potash is employed, and, since hot potash rapidly destroys glucose, the liquid product must first be thoroughly cooled, and precaution taken to add the potash slowly so as to avoid any marked rise of temperature. The liquid is then brought up with water to a given bulk and filtered through a dry filter. It frequently happens that more or less colour appears as the result of neutralisation, and sometimes, after standing, a certain amount of coloured matter separates out, leaving, on filtration, the liquid in a much better condition for examination.

By treatment of the coagulated residue from the alcoholic extraction, the carbohydrate other than sugar—carbohydrate of an amylose nature—is obtained, and the process to be adopted is the same as that resorted to for obtaining and estimating the glycogen of the liver, viz., dissolving by means of potash, pouring into spirit, collecting the precipitate, and subjecting it to the inverting action of sulphuric acid.

The residue is placed in a flask together with 50 c.c. of a 10 per cent. solution of potash. It is important that the material should have been as thoroughly as possible reduced to a minute state of subdivision, in order to secure that none of the nitrogenous matter escapes the action of the potash and remains in a state to give rise, after the subsequent treatment with acid for conversion into glucose, to the biuret reaction when submitted to titration with the copper solution. It is of advantage to allow the mixture to stand for twenty-four hours in the cold, as greater security is thus given of complete solution being effected by the thirty minutes' boiling subsequently resorted to.

To prevent concentration, through evaporation, of the potash solution, and the possible destruction of the carbohydrate in the process of boiling, the operation should be performed in a flask fitted to an inverted condenser, and, in order to escape from the inconvenience arising from the frothing which not unfrequently occurs, a large-sized flask is required.

The 50 c.c. of liquid are now poured into about 500 c.c. of spirit to separate and precipitate the carbohydrate present. With a smaller proportional amount of spirit, a risk is incurred of the precipitation being incomplete. By setting the beaker aside until the following day, the thorough settlement of the precipitate is promoted, and in this way the subsequent process of separation by filtration facilitated.

For filtration, a plug of glass-wool should preferably be used. With the employment of filter paper, which is not only constituted of cellulose, but is, moreover, sometimes found to contain starch, a risk of error is incurred through the possibility of glucose-forming matter being introduced in the subsequent removal, or washing off, of the precipitate.

The precipitate on the glass-wool is next thoroughly washed with spirit, and the glass-wool plug then turned into the beaker, and the funnel washed with a little hot water. If much water should have been required for washing off the precipitate, evaporation should be resorted to to bring the bulk down to about 50 c.c. The product, after its transference to a suitable-sized flask, and the addition of sulphuric acid to the extent of 2 per cent., is submitted to the action of heat, according to the directions given further on, for effecting conversion into a cupric oxide reducing sugar.

A certain amount of undissolved black matter is usually afterwards noticeable, and it is advisable that this should be separated by filtration, before proceeding to neutralisation. Otherwise, it may be taken up by the potash employed, and interfere with subsequent titration by giving a more deeply coloured product. The filter used should be small enough to admit of being thoroughly washed without unduly increasing the bulk of the liquid. The acid is then neutralised with potash, cautiously added. This leads to a further separation of matter in the form of a dark-coloured, and sometimes bulky, sediment, which should be removed by filtration through a dry filter after the liquid has been made up with water to a definite volume.

The amount of glucose present is now determined by titration with the ammoniated cupric test, and the result may be either represented as carbohydrate expressed as glucose or brought by calculation into glycogen—amylose carbohydrate—figures. The equivalents of glucose and the amylose group of carbohydrates corresponding with their accepted formulæ—$C_6H_{12}O_6$ and $C_6H_{10}O_5$—are respectively 180 and 162, giving the fraction 162/180, or 0·9, as the multiplier for the conversion of glucose figures into glycogen figures.

Such is the method of dealing with a fluid animal product like blood. If a solid organ or tissue, such as liver, muscle, &c., is being examined, it is first thoroughly pounded in a mortar, or otherwise minutely divided, and a weighed portion placed in a relatively large quantity of alcohol and well stirred up. After standing for twenty-four hours, the alcohol is strained off through muslin, the residue is pressed, and the extraction and pressing twice repeated in the manner already described. By this process, the sugar present is thoroughly extracted, and separated from the other carbohydrate matter, which remains behind in the residue. The alcoholic extract and the residue are then treated as in the case of blood, for procuring suitable products for examination.

Until recent investigation taught me otherwise, I was under the impression that the carbohydrate matter obtained by treatment of the residue with potash consisted simply of glycogen. Where glycogen is present, as in the case of the liver, muscle, and certain other constituents of the organism, it enters, it is true, to a greater or less extent into the result; but, as I have shown when speaking of the glucoside constitution of proteid matter, the effect of the action of potash upon proteid is to split off from it carbohydrate material, which passes with the glycogen that may be present, and which, when none is present, I formerly took erroneously to be glycogen.

The carbohydrate of proteid origin, as I have previously mentioned (pp. 36, 37), offers much greater resistance than glycogen to the converting action of sulphuric acid. The 2 per cent. strength of acid, which suffices readily to carry glycogen fully into glucose, fails with the body in question to do more than give it about half the amount of cupric oxide reducing power that is given to it by acid of 10 per cent. strength, and even with acid of this latter strength I consider it doubtful whether complete conversion into glucose is effected. It hence happens, in the case of this cleavage carbohydrate, that if the reckoning be applied to it that is applicable in the case of glycogen, an under-estimation of material will result to the extent of about half or, it may be, even more.

A difficulty is obviously created by this cleavage carbohydrate in the estimation of glycogen as a constituent of the components of the body. As, however, the amount of the cleavage material, given after the employment of 2 per cent. sulphuric acid as representing glucose, may be computed to stand at not more than about 2 or 3 per 1000 of the fresh tissue examined, it may be considered that where glycogen is present to any significant extent, as may be looked for to be the case with the liver and some other structures, the error introduced does not amount to anything of consequence.

A point of paramount importance, in connexion with the employment of sulphuric acid for conversion of glycogen, maltose, &c., into glucose, is the necessity of attention being given to secure that the sulphuric acid added remains as free acid in the solution, and is not appropriated in neutralising an alkali, or in displacing a weaker acid, as where acetic or citric acid has been used and afterwards neutralised in some earlier stage of the process.

As regards the length of time to be employed for conversion, it may be stated that in operating upon starch, glycogen, dextrin, or maltose, it is advisable, in order to be quite safe that full conversion has taken place, to boil for an hour and a half, though probably a shorter time may in reality suffice. An inverted condenser must be used to avoid concentration of the acid liquid, and thereby risk of destruction of the product.

The following observations may be introduced to show the extent of transformation met with after boiling with 2 per cent. acid for successive lengths of time.

A decoction of starch boiled with sulphuric acid (2 per cent. strength) with the employment of the inverted condenser, and 10 c.c. removed and analysed at the undermentioned periods:—

At the end of	Cupric oxide reducing power, expressed as glucose.	
	Obs. 1.	Obs. 2.
$\frac{1}{4}$ hour..........	0·200 gram.	0·140 gram.
$\frac{1}{2}$,, ...	0·380 ,,	0·248 ,,
$\frac{3}{4}$,, ...	0·620 ,,	0·264 ,,
1 ,, ...	0·660 ,,	0·328 ,,
$1\frac{1}{2}$,, ...	0·660 ,,	0·350 ,,
2 ,, ...	0·666 ,,	0·350 ,,
3 ,, ...	0·660 ,,	0·350 ,,

A solution of maltose, derived from the action of saliva upon starch, similarly boiled with sulphuric acid, and 10 c.c. removed and analysed at the undermentioned periods:—

At the end of	Cupric oxide reducing power, expressed as glucose.
5 minutes	0·025 gram.
10 ,, ...	0·031 ,,
15 ,, ...	0·031 ,,
30 ,, ...	0·034 ,,
45 ,, ...	0·041 ,,
1 hour	0·041 ,,
$1\frac{1}{2}$,, ...	0·041 ,,

On looking at the above results, it is observable that in the last experiment conversion was completely effected within three-quarters of an hour, whilst in the first two a longer time elapsed before the complete change was found to have occurred.

With exposure to a higher temperature, a comparatively short time suffices for the complete conversion of starch, glycogen, &c., into glucose. Taking advantage of this fact, I am in the habit of using an autoclave—a boiler constructed for the application of heat under pressure. When used for the purpose referred to, it is set at a pressure equivalent to five atmospheres, which gives a temperature of about 150° C. (300° F.). At this point, the conversion into glucose is accomplished within about a quarter of an hour. It is reliably secured by half an hour, which is the length of time I consider it advisable to allow.

It is requisite to mention that the presence of sodium sulphate interferes with the inverting action of the acid when the process is conducted under ordinary atmospheric pressure, and that error may thus arise where the inverted condenser is used. Under the higher pressure, however, at which the autoclave is worked, this interference is not, as far as I have perceived, exerted in such a manner as to be noticeable.

In the case of lactose, one hour suffices for complete conversion into glucose, and may be taken as the time to be allowed. The inverted condenser must be employed, to prevent concentration by evaporation. The use of the autoclave is inadmissible, on account of its leading to the development of colour in the product.

With cane sugar, inversion is accomplished within a few minutes. On account, however, of the charring action exerted by sulphuric acid upon cane sugar, citric acid is a preferable agent to use for its inversion. This acid is equally efficacious for the purpose required, and its employment is not attended with the risk of any destruction of carbohydrate. Boiling for seven minutes with a citric acid solution of 2 per cent. strength amply suffices for securing full conversion into glucose. The operation may be conveniently performed in a small open flask, the employment of an inverted condenser being unnecessary.

On account of the greater proneness of cane sugar and its derivative invert sugar to undergo destruction by the action of sulphuric acid than other carbohydrates, the use of the autoclave should be avoided where the presence of these sugars is suspected.

Qualitative and Quantitative Testing.

Having described the processes by which the carbohydrates existing in animal products are separated from other matters and prepared for the application of testing, I have next to deal with the manner in which the process of testing is conducted.

The copper test, based on the power possessed by glucose and some other sugars of reducing cupric oxide (CuO) to the state of cuprous oxide (Cu_2O), affords, I consider, by far the most suitable means of obtaining the information desired, whether of a qualitative or quantitative nature.

The form of copper solution ordinarily employed is a liquid in which cupric oxide is held in solution by alkali through the influence of an alkaline tartrate. The test perhaps in most general use is the liquid known as Fehling's solution, which has the following composition:—

Fehling's Solution.

Cupric sulphate .... 34·65 grams.
Potassic sodic tartrate (Rochelle salt) 173·00 ,,
Caustic soda solution (sp. gr. 1·1200) 600·00 c.c.
Distilled water to 1 litre.

Fehling's solution is, in my opinion, improved by using a larger proportion of alkali than that commonly employed. Thus modified, it is more stable and less liable to give rise to fallacious indications by undergoing spontaneous reduction on boiling. I am accustomed also, on grounds of convenience, to use potash instead of soda. The following is the composition of the liquid I employ:—

Modified Fehling's Solution.

Cupric sulphate .... 34·65 grams.
Potassic sodic tartrate (Rochelle salt) 170·00 ,,
Caustic potash.................... 170·00 ,,
Distilled water to 1 litre.

It is necessary to bear in mind that these cupric solutions, the latter, however, less readily than the former, undergo, on keeping for some time, especially under exposure to light, change of such a nature as to lead to the spontaneous deposition of some suboxide, and to the deposition of more on boiling. When in a state thus to give rise to the appearance of suboxide precipitate on boiling, the liquid may be restored for a while to its former condition by the addition of alkali.

Due attention must needs be given to the source of fallacy here presented, and when this is done the test may be regarded as an exceedingly valuable one for the recognition of sugar.

The test may also be made use of for *quantitative* purposes, and be thus applied either gravimetrically or volumetrically. Its employment for gravimetric determination is based on the fact that five molecules of cupric oxide are reduced by one molecule of glucose; and, for volumetric determination, on the fact that the amount of cupric oxide contained in 10 c.c. of the liquid is just reduced by 0·050 gram of glucose.

For the *gravimetric* process, the product to be examined is boiled with an excess of the cupric solution. In the case of blood, &c., the preparation of the product may be conducted by the sodium sulphate method described at p. 59. After full reduction has taken place, the cuprous oxide is collected and may be weighed as such, and the sugar calculated upon the basis of one molecule of glucose for every five molecules of cuprous oxide obtained.

A better method, on account of its extreme delicacy and precision of dealing with the precipitated suboxide, consists in dissolving it in nitric acid and then submitting the solution to galvanic action in such a manner as to occasion the deposition of the whole of the copper in a pure metallic state upon a weighed cylinder of platinum foil. The weight of the copper is thus readily ascertained and the required calculation made. This process, which I at one time employed when engaged upon a series of comparative observations on the amount of sugar existing in blood taken from different parts of the system, is described in detail in a communication presented by me to the Royal Society in 1877, and contained in the 'Proceedings' for that year (Vol. 26, p. 314). The results which will be introduced later on will afford evidence of the exactness of the process as a means of sugar estimation.

Holding the position it does, as regards reliability, the electrogravimetric process is not only of intrinsic value for the purpose of quantitative analysis, but may be considered to possess additional value as a means of gauging the accuracy of other tests. It is, however, a lengthy and, it must be said, a tedious process, and is, on this account, not well adapted for ordinary employment.

The estimation of sugar gravimetrically through the copper test, by whatever method conducted, is open to the important source of fallacy that if ammonia (whether free or in combination), or a nitrogenous product capable of developing it during the process of boiling with the fixed alkali of the test, is present, error will necessarily follow, in consequence of the solvent action it exerts upon the cuprous oxide, and its effect in thereby leading to more or less of this product escaping deposition and thus failing to appear in the result obtained. It may be assumed, in the case of the liver, that this in reality occurs, for the figures yielded by the method are so absolutely insignificant and so inconsistently low, compared with those obtained in other ways, that it is evident they cannot be correct. With blood, in a fresh state, I have no reason to think that any such source of fallacy exists. Indeed, the evidence that is presented points to the contrary.

In the *volumetric* method, the estimation is effected by noting the fading and loss of the blue colour of the test solution, as the conversion of its cupric oxide into cuprous oxide takes place. The method is a much more expeditious one than the gravimetric, and is not influenced detrimentally by the presence of ammonia. The reckoning is based upon the cupric oxide contained in 10 c.c. being just reduced by 0·050 gram of glucose, which is equivalent to saying that the blue colour belonging to 10 c.c. of the solution is just removed by 0·050 gram of glucose. Theoretically, there is nothing to be said against this process; but, practically, it is found that its application is attended with the disadvantage of the reduced oxide interfering, by remaining suspended in the liquid, with the accurate recognition of the terminal point of reduction. For purposes where minute accuracy is not essential, a sufficiently approximate result can be obtained, but for physiological purposes, and in other cases where great precision is required, the process is unsuited for employment.

Quantitative Determination of Sugar by the Ammoniated Cupric Test.

A test which I introduced about fifteen years ago, and of which a daily experience during the whole of this period enables me to speak with confidence, is not open to the objection above alluded to, and has, moreover, special advantages in other respects, rendering it eminently valuable as an agent for sugar estimation. I refer to the ammoniated cupric test, the principle of action of which is that through the presence of ammonia the cuprous oxide formed by the reducing influence of sugar is held in solution in a colourless state, instead of appearing as a coloured deposit. The progress of the reduction is thus attended with a simple fading of colour, unobscured throughout, and no difficulty is experienced in recognising precisely when the terminal point is attained.

The ammonia present in the liquid, whilst holding in solution the reduced oxide of copper, does not interfere with the process of reduction. It is incidentally advantageous in deepening the colour of the cupric solution, and thereby increasing the defining capacity of the test, which is proportional to the range of colour passed through. Further—and this is a point of great importance—ammonia confers upon the solution a perfect self-preservative power, the conditions being such that so long as any air is in contact with the fluid, the copper is of necessity kept in a fully oxidised state, and prevented from undergoing deposition.

The test is an exceedingly delicate one, and is further recommended by its readiness and speediness of application. It possesses, indeed, all the delicacy and reliability of the electro-gravimetric process, together with more general applicability, and infinitely greater facility for use. When the two processes are applied side by side, the results yielded are found to stand in the closest agreement. It is not too much to say that without the assistance of the ammoniated cupric test I could not have obtained the information which stands at the foundation of this work.

The following is a description of the mode of preparation of the test liquid, with particulars regarding its position in relation to cupric oxide reduction and sugar oxidation.

I originally directed that the ammoniated cupric test should be prepared from Fehling's solution, by adding to 100 c.c. of it 300 c.c.

of ammonia and 600 c.c. of distilled water; and, in operating with the liquid, I at first took it for granted that the same relation existed between the amount of oxide of copper reduced and that of sugar oxidised, as under the employment of the copper test of the ordinary, or non-ammoniated, kind, namely, that five molecules of oxide of copper were reduced by one molecule of glucose. The liquid made in this way contained one-tenth of Fehling's solution, and if it comported itself in the same manner as the latter, 10 c.c. of it would stand equivalent to $0\cdot005$ gram of glucose. In working with it, the results obtained stood in harmonious relation with each other, but when checked by trying it with liquids containing known quantities of sugar it was found that the figures given were invariably too high. At first, I was at a loss to account for this result, but subsequent observation showed that in the case of the ammoniated liquid six molecules of cupric oxide, instead of only five as with Fehling's solution, are reduced by one molecule of sugar. When the reckoning was made upon this basis, the results exactly corresponded with the actual amounts of sugar known to be present. Moreover, with liquids containing glucose, examined comparatively with Fehling's solution and the ammoniated cupric test containing one-tenth of Fehling's solution, the results stood exactly in accord, under the reckoning that five molecules of cupric oxide in the one case and six molecules in the other were reduced by one molecule of glucose.

It was further revealed, in the course of the enquiry conducted, that, with the liquid prepared as above from Fehling's solution, the result of titration was to some extent dependent upon the manner in which the product was dropped in from the burette. When the rate of dropping was such as not to involve a prolonged period of boiling, a correct result was obtained, but when the operation was more slowly performed, evidence was afforded of the occurrence of a certain amount of spontaneous reduction. This is comprehensible from the behaviour of Fehling's solution itself. When in a fresh and not over-dilute state, it does not, even on prolonged boiling, show any sign of the occurrence of reduction, but if brought to a very dilute state, it undergoes, on ebullition, spontaneous change attended with a deposition of suboxide. The ammoniated liquid as compared with Fehling's solution presents considerable dilution. It is, therefore, not surprising that it should be unable to resist the influence of unduly prolonged ebullition without undergoing spontaneous change.

In experimenting with the test, it was found that the addition of fixed alkali to the extent of 1 gram of potash for every 20 c.c. had the effect of removing the instability referred to, and of giving to the liquid the power of resisting the tendency to spontaneous reduction under exposure to a prolonged period of boiling. Further investigation showed that a great excess of fixed alkali, such, for instance, as resulted from the addition of 5 grams of potash for every 20 c.c., modified the character of the liquid, and brought it into the same position as the ordinary copper solution, in relation to the number of molecules of cupric oxide reduced by one molecule of sugar; in other words, that whilst the addition of potash in the proportion of 1 gram for every 20 c.c., and somewhat over this, did not modify the character of the liquid in relation to extent of reduction occurring, the addition of 5 grams to the 20 c.c. so far altered the behaviour of the test that five molecules only of cupric oxide, instead of six, were reduced by one molecule of glucose. With the addition of potash in quantities somewhat under this, the amount of cupric oxide reduced by one molecule of glucose was found to stand between the five and the six molecules.

To bring the ammoniated copper liquid into a position to represent an exact tenth of the sugar-value of Fehling's solution, the proportion of copper requires to be increased so as to give six molecules against five. When this has been done, as by taking 120 c.c. of Fehling's solution instead of 100 for the litre of the ammoniated liquid, the results obtained from titration coincide for the two liquids, and conform with those that are to be looked for when known quantities of sugar, determined by the balance, are employed.

When Fehling's solution is used for the preparation of the ammoniated liquid, the required stability, without alteration of sugar-value, is given by doubling the amount of caustic soda present. With the modified Fehling's solution, the composition of which is given at p. 68, and which, for all purposes, I consider a preferable liquid for use, the amount of fixed alkali is sufficient to meet what is wanted.

No need exists, however, to prepare the ammoniated cupric test from either of these solutions, and I never, now, have recourse to their employment for the purpose. I consider it better to proceed in a direct way, and the following are the quantities to be used :—

Ammoniated Cupric Test Solution. (Pavy.)

Crystallised copper sulphate	4·158 grams.
Potassic sodic tartrate (Rochelle salt) ..	20·400 ,,
Caustic potash	20·400 ,,
Strong ammonia solution (sp. gr. 0·880)	300 c.c.
Distilled water to 1 litre.	

In preparing the test, it is found desirable to dissolve the potash and Rochelle salt together in a portion of the water, and the copper sulphate separately in another. The latter requires the aid of heat for solution. When completely dissolved, it is poured into the mixture of potash and Rochelle salt.

After the mixed liquids have cooled, the ammonia is added, and then a sufficiency of water to bring up the volume to a litre.

In sugar-value, 10 c.c. of the liquid stand equivalent to 0·005 gram of glucose.

Such is the theoretical sugar-value of the solution, from the copper present. For many purposes it may be accepted as the actual value, but when security of close accuracy is required the liquid should be standardised by means of a solution containing a known quantity by weight of sugar. This I am in the invariable habit of doing when the test is used in conducting physiological investigations. For the purpose, it is best to employ cane-sugar and to convert it into glucose by the agency of an acid. What is sold in the form of coarse colourless crystals, under the name of "white crystal," and used particularly for sweetening coffee, constitutes a very pure kind of cane-sugar, and therefore an advantageous one to take. It should be reduced to a fine powder, and freed from adherent moisture by exposure for a short time in a hot-air oven to a temperature near 100° C. (212° F.). Afterwards it may be kept over sulphuric acid in a desiccator, ready for use at any time. A carefully weighed portion, say 0·250 gram, is subjected to inversion by boiling for seven minutes with 50 c.c. of 2 per cent. citric acid. This acid is employed in preference to a mineral acid, to avoid any chance of destruction of sugar. When cool, the acid is neutralised with potash, and the volume made up with distilled water to 250 c.c. The liquid is subsequently titrated with the ammoniated solution, and the value of the test calculated from the result obtained. No disparity of any account should be found between the theoretical and the actual value.

In the employment of the test solution, 10 c.c. and 5 c.c. are convenient quantities to work with, and do not call for the use of any appliance to obviate the inconvenience arising from the evolved ammonia, such as I found to be necessary when I first introduced the test and used a larger quantity for operating with. By diluting the specimen to be examined, as much accuracy is attainable as with the employment of a larger amount of the test and a less dilute product.

The apparatus best suited for the application of the test consists of a flask of about 150 c.c. capacity, through the cork of which passes a delivery tube from a Mohr's burette, for dropping in the product to be examined, and also an exit tube for the escape of air and steam. The burette is graduated into 20 c.c., each centimeter being divided into tenths, and is fixed in a suitable stand (*vide* accompanying photo-engraving). Instead of the spring clip commonly used to regulate the dropping of fluid, I employ a screw arrangement, by which the flow is susceptible of being regulated to a nicety.

76 ANALYTICAL STEPS OF PROCEDURE.

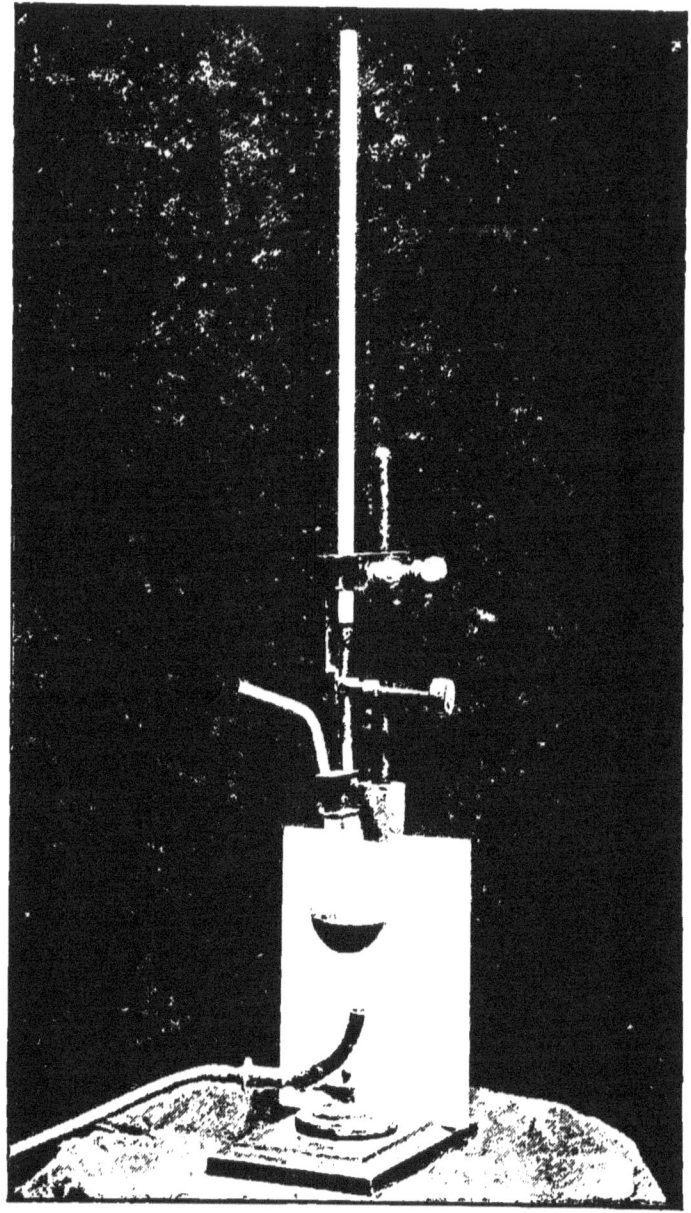

The product for examination having been introduced into the burette, a little is allowed to escape in order to free the delivery tube from air-bubbles, after which the height of the liquid is read off and noted down. The 10 c.c. or 5 c.c. of test liquid, or whatever quantity it is intended to employ, are then placed in the flask, and about 20 c.c. of distilled water added to increase the bulk to a convenient extent for boiling to be performed. The flask is now affixed to the cork belonging to the burette and allowed to hang suspended over the flame of a spirit lamp or Bunsen burner. To facilitate observation of the progress of decoloration, a white background of opal glass is provided. When the contents of the flask have well commenced to boil, the screw which governs the flow from the burette is turned so as to permit the liquid to escape by drops at the rate of about 60 to 100 per minute, according to the effect produced upon the colour of the test. What is wanted is a gradually advancing decoloration until the contents of the flask are brought to a perfectly colourless state. Towards the end, the dropping in has to be conducted more slowly than at first, so as to avoid going beyond the exact point required. When decoloration is complete, the flow is stopped. A reading of the level of the liquid in the burette shows the amount of the product which has been used in decolorising the 10 c.c. or 5 c.c. of test solution; in other words, the amount which contains sugar equivalent in cupric oxide reducing power to 0·005 gram or 0·0025 gram, as the case may be, of glucose.

A basis is thus afforded for ascertaining the amount of sugar contained in the entire product, and, as the product represents the weighed material taken for examination, calculation suffices to give a per cent. or per mille expression of the amount of sugar that was present.

If, in conducting the analysis, the contents of the burette are dropped in too slowly and the boiling becomes prolonged, some suboxide may fall in consequence of the expulsion of the ammonia before the operation is completed. Should this occur, a fresh titration must be undertaken, and the contents of the burette dropped in a little more quickly. Any suboxide happening to have been deposited upon the surface of the flask in a previous titration promotes a more speedy deposition of suboxide than would otherwise occur. It must, therefore, be removed by a little nitric or other mineral acid before the flask is again used.

The chief precautions to be observed in conducting a titration are to avoid dropping in the liquid from the burette so rapidly as to incur the risk of running beyond the point required for decoloration, and, at the same time, not to drop in so slowly as to lead to a deposition of suboxide from the expulsion of the ammonia. The liquid should be kept steadily boiling, so that the upper part of the flask may be continuously filled with vapour. An interruption of ebullition would permit the entrance of air, and thereby some possible re-oxidation of the reduced oxide, which would have to be again reduced before the end of the process, thus leading to a fallacious result.

With everything conveniently at hand, a few minutes only are required for the performance of the process, from beginning to end. To give precision to the analysis, at least two, and sometimes three or four, titrations are required to be conducted. If the first two titrations yield accordant results, they may be considered to suffice. The first titration, however, is obviously of a somewhat tentative nature, in the absence of any previous knowledge of the quantity of liquid that will be required to be dropped in from the burette. The titrations should be repeated until closely accordant results are obtained, the mean of which may be accepted as the final result.

In first employing the test, the operator must be prepared to find that a certain amount of practice is required to work with it satisfactorily. When, however, the requisite experience has been gained, no difficulty is met with in obtaining precise results. Proof is easily afforded of the precision of which the test is susceptible by examining two liquids containing different amounts of sugar, and then mixing them in equal proportions and examining the mixture. The figures given by the latter, under experienced hands, and when the liquids used are in a good, colourless state, will be found to exactly, or almost exactly, coincide with the calculated mean drawn from the separate examinations previously conducted.

As a further illustration of the precision attainable through the medium of the test, I may adduce a few results of duplicate analyses belonging to experiments recently conducted at the Examination Hall Research Laboratories for the purpose of extending and checking my former work. The results in question serve, it may be remarked, not only to illustrate what has been mentioned with regard to the test, but also to show the satisfactory position in which the analytical procedure, taken in its entirety, stands.

Duplicate Analyses with the employment of the Ammoniated Cupric Test.

	Sugar per 1000, expressed as glucose. Duplicates.	
I. Dog's blood from right ventricle—		
Before sulphuric acid	0·960	0·940
After ,,	0·920	0·937
II. Dog's blood from portal vein—		
Before sulphuric acid	0·977	0·943
After ,,	1·100	1·070
III. Dog's blood from portal vein—		
Before sulphuric acid	0·968	0·897
After ,,	1·277	1·130
IV. Dog's blood from portal vein—		
Before sulphuric acid	1·823	1·723
After ,,	2·426	2·187
V. Rabbit's blood from portal vein—		
Before sulphuric acid	1·575	1·630
After ,,	2·085	2·085
VI. Rabbit's muscle—		
Before sulphuric acid	2·130	2·297
After ,,	3·640	3·743
VII. Rabbit's muscle—		
Before sulphuric acid	2·427	2·570
After ,,	3·640	3·493

Evidence of a somewhat different nature is also adducible in support of the validity of the results yielded by the process of analysis that has been described. Some years ago I subjected specimens of blood to examination by the electro-gravimetric process, and also by the ammoniated cupric volumetric process. The results obtained are recorded in the 'Proceedings of the Royal Society' for April, 1879, and, although the two processes for the determination of the sugar differ in principle, the one being dependent upon the deposition and collection of the suboxide of copper, and the other upon the reduction being measured by the loss of blue colour accompanying the transformation of the cupric into cuprous oxide, the figures obtained, as will be seen by the subjoined table, present so close a general accord as to mutually confirm the trustworthiness of each other.

Results from the Analysis of Blood by the Gravimetric and the Ammoniated Cupric Volumetric Processes.

Sugar per 1000, expressed as glucose.

	Gravimetric process. Mean of two analyses.	Ammoniated cupric volumetric process.
I. Sheep	0·589	0·571
II. Bullock	0·735	0·650
III. Bullock	0·921	0·896
IV. Sheep	0·533	0·567
V. Bullock	0·511	0·559
VI. Sheep	0·631	0·650

INGESTED CARBOHYDRATES TRACED TO THE PORTAL BLOOD.

We have next, by the aid of the methods described in the foregoing pages, to follow the main carbohydrates concerned in alimentation, from their introduction into the mouth to their passage, by absorption, into the blood of the portal system, and see what changes occur. Attention will be given to them in the following order:—

Starch, with its congener glycogen and their derivatives, dextrin and maltose.
Cellulose.
Cane sugar.
Lactose.
Glucose.

Starch, &c.

Starch enters more largely than any other carbohydrate into the composition of the food of animals, and occupies, on this account, the position of chief importance. Although itself exclusively derived from the vegetable kingdom, it is represented in the animal kingdom by its congener glycogen. Glycogen, so far as is to be learnt from the indications of cupric oxide reducing action, comports itself in its transformations precisely in the same manner as starch, and, since it only occurs to a comparatively insignificant extent as a constituent of food—chiefly in muscular tissue and liver—it will not receive separate attention in the succeeding account of the digestive changes, but must be understood to be included in what is said under the head of starch.

As has been previously mentioned, starch exists naturally in the form of a granule, composed of successive layers, of which the outer layer offers great resistance to solvent influences. When the granules are ruptured, however, by the application of heat, an aqueous decoction is obtainable; but, even in this form, the starch, by virtue of its colloidal properties, is unsusceptible of being absorbed from the alimentary canal. To be serviceable, therefore, as an alimentary article, starch must, as a necessary preliminary to absorption, be brought into a soluble and diffusible form, and this is effected by the process of digestion. Through the influence exerted upon starch in the digestive system, it becomes converted into a form of sugar, with an admixture of the intermediate products of transformation known as dextrins.

There are various secretions possessing this transformative power over starch—secretions, that is to say, containing an amylolytic ferment. The first with which the food comes into contact is the saliva, a somewhat viscid, slightly alkaline secretion, containing an unorganised ferment known as ptyalin. The extent of the change exerted by saliva is probably small, owing to the incompleteness, as well as the short duration, of the contact between the ferment and the starch.

Experiments with starch and saliva show that maltose, and not, as was formerly believed, glucose, is the main end product of the action of ptyalin. Apparently, however, a certain, but insignificant, amount of glucose is produced at the same time. As to the precise nature of the changes occurring, and of the intermediate products formed, authorities are at variance, and it is not necessary to enter into the question of the theoretical mode of splitting up of the starch molecule further than has been done in a former part of this work. Broadly speaking, the change may be described as one of increasing hydration, attended with advancing cupric oxide reducing power.

The action may be studied, and information obtained regarding the general nature of the products, by adding to a decoction of starch a little saliva, and making observations on successive portions taken at short intervals. By means of alcohol the products of transformation may be obtained free from any admixture of untransformed starch, should such exist; and by subsequent treatment with the ammoniated cupric liquid, before and after boiling with sulphuric acid, the cupric oxide reducing power, as a simple expression, is obtained. In eleven observations that I find recorded in my laboratory books on mixtures of saliva and starch solution, the cupric oxide reducing power of the respective products varied from 39 to 60, taken in relation to that of glucose reckoned as 100. Thus, in certain cases, conversion of the starch into maltose had practically been accomplished, whilst in others a large proportion of dextrin, or something with a lower cupric oxide reducing power than that of maltose, was still present.

After the short period of detention that the food undergoes in the mouth, it is propelled along the gullet to the stomach, there to be brought into contact with the gastric juice.

Acids, as is well known, exert an influence in the direction of checking the amylolytic action of saliva; and it is generally asserted that, on the arrival of starch in the stomach, this action is at once stopped by the hydrochloric acid existing in the gastric juice. This statement is not, however, to be accepted too literally or too absolutely, for it must be remembered that the gastric juice is not normally produced except in response to the stimulus of the presence in the stomach of extraneous matter, so that food received into a stomach previously empty will not immediately come under the influence of this secretion; and, during the short interval, whatever may be its duration, which elapses between the entry of food into the organ and the subsequent flow of gastric juice, starchy matters may undergo some further change under the continued action of the saliva. Beyond the interference with salivary action, gastric juice exerts no influence upon starch digestion.

A power of absorption of diffusible principles is possessed by the lining membrane of the stomach, and, by virtue of this power, whatever diffusible matter may have been produced by salivary action stands in a position to pass, at this early stage, into the circulation. That such matter is present in considerable amount in the contents of the stomach, after the ingestion of starchy food, is shown by observation. In illustration, I may refer to the results obtained from an examination of the stomach contents of ten rabbits, killed at a period of digestion. In each instance, an alcoholic extract was made, for the purpose of obtaining the transformed carbohydrate matter free from admixture with starch. The examination of the extracts revealed the presence of a considerable amount of carbohydrate matter possessed of cupric oxide reducing power, varying in the several instances from 25 to 53 expressed in relation to that of glucose taken at 100. In two of the observations, a portion of the contents of the stomach was dealt with at once, whilst another portion was allowed to stand for thirty minutes at a temperature of 48° C., previous to extraction with alcohol. Upon examination, it was found that the cupric oxide reducing power had become raised in the one case from 25 to 48, and in the other from 25 to 50. It thus appears, be the precise cause what it may, that a certain amount of change of carbohydrate may occur under certain conditions, within the stomach.

From the stomach the alimentary matter, in a semi-fluid condition, is conducted to the small intestine, the part of the alimentary tract which may be regarded as constituting the main seat of starch digestion. It is here brought under the influence of the bile, the pancreatic juice, and the secretions derived from the intestinal walls. Bile, through the alkali contained in it, contributes towards neutralising the acidity of the chyme as it passes from the stomach, and thus, indirectly assists in promoting the subsequent digestion of starch by the other secretions; otherwise little or no action is probably exerted by it upon the carbohydrates.

Pancreatic juice is an alkaline secretion, possessed of marked power of acting upon starch, its action, like that of saliva, leading to the production of dextrins and maltose. Within the intestinal canal there exist the most favourable conditions for the exercise of the transformative power—for example, intimate contact, alkalinity, and a moderately elevated temperature—and, accordingly, the conversion of starch begun in the mouth is now carried on with its highest degree of activity. The process of transformation and the nature of the products formed may be studied in a similar manner to that mentioned in the case of saliva. The results of ten experiments with starch and pancreatic ferment derived from the sheep, the cat, and the dog, showed the attainment of cupric oxide reducing power ranging from 23 to 60, expressed in relation to glucose taken at 100. Thus in some cases complete conversion into maltose had practically been accomplished, whilst in others the product must have consisted largely of dextrin.

The effect of the presence of acid or of carbonated alkali is to modify the action of the amylolytic ferments, whether pancreatic or salivary. For instance, the addition of sodium carbonate seems to determine the formation of a final product with a lower cupric oxide reducing power than that which would otherwise be obtained. The change is prevented from passing beyond a certain point more or less short, apparently according to the amount of sodium carbonate used, of the production of maltose. This is shown by the subjoined observations.

100 c.c. of a decoction of starch were mixed with 50 c.c. of an aqueous extract of pancreas, and exposed to a temperature of 48° C. 10 c.c. portions were removed at the undermentioned periods, and the cupric oxide reducing power of each taken. Parallel experiments were made with the addition of 2 grams and 4 grams (giving the respective percentages of 1·3 and 2·6) of carbonate of soda. The cupric oxide reducing power, expressed as glucose, was found in the several instances to stand as follows:—

10 c.c. portions after	Sugar found, expressed as glucose.		
	Without Na_2CO_3.	With 1·3 per cent. of Na_2CO_3.	With 2·6 per cent. of Na_2CO_3.
15 minutes	0·015 gram	0·006 gram	0·005 gram
30 ,,	0·018 ,,	0·009 ,,	0·009 ,,
45 ,,	0·018 ,,	0·009 ,,	0·009 ,,
1 hour	0·018 ,,	0·009 ,,	0·008 ,,
2 hours	—	0·009 ,,	0·009 ,,

A similar experiment was upon another occasion performed with a different decoction of starch and aqueous extract of pancreas. The following, expressed as before, were the results obtained:—

10 c.c. portions after	Sugar found, expressed as glucose.		
	Without Na_2CO_3.	With 1·3 per cent. of Na_2CO_3.	With 2·6. per cent. of Na_2CO_3.
15 minutes	0·031 gram	0·013 gram	0·009 gram
30 ,,	0·046 ,,	0·014 ,,	0·010 ,,
45 ,,	0·047 ,,	0·014 ,,	0·010 ,,
1 hour	0·046 ,,	0·014 ,,	0·010 ,,

Acids, as already mentioned with reference to saliva, exert a retarding or an actually preventive influence on the action of these ferments upon starch. The presence of even an organic acid, such as citric, whilst not wholly preventing the transformation, exerts a marked degree of interference with such change, and determines the non-conversion of a considerable quantity of starch.

The following results of experiments made with starch solution and aqueous extract of the pancreas of a dog, will serve to show the influence exerted by citric acid and sodium carbonate respectively on the process of transformation.

30 c.c. of a solution of starch were placed in contact with 5 c.c. of the pancreatic extract and allowed to stand for 18 hours at the ordinary temperature. At the end of that time an alcoholic extract was made, and the sugar determined before and after treatment with sulphuric acid. A parallel procedure was adopted with the addition, in the one case, of 0·5 gram (1·4 per cent.) sodium carbonate, and, in the other, the same quantity of citric acid. The results obtained stood as follows:—

	Sugar produced, expressed as glucose in grams.	Relation of the cupric oxide reducing power of the product to that of glucose at 100.

Experiment I.

After action of ferment alone	{ before sulphuric acid .. { after ,, ,, ..	0·074 0·156	} 47
After action of ferment in presence of 1·4 per cent. sodium carbonate	{ before sulphuric acid .. { after ,, ,, ..	0·034 0·130	} 26
After action of ferment in presence of 1·4 per cent. of citric acid	{ before sulphuric acid .. { after ,, ,, ..	trace 0·058	} —

Experiment II.

After action of ferment alone	{ before sulphuric acid .. { after ,, ,, ..	0·070 0·138	} 51
After action of ferment in presence of 1·4 per cent. of sodium carbonate	{ before sulphuric acid .. { after ,, ,, ..	0·034 0·148	} 23
After action of ferment in presence of 1·4 per cent. of citric acid	{ before sulphuric acid .. { after ,, ,, ..	0·032 0·096	} 33

		Sugar produced, expressed as glucose in grams.	Relation of the cupric oxide reducing power of the product to that of glucose at 100.

Experiment III.

After action of ferment alone	before sulphuric acid .. after ,, ,, ..	0·074 0·322	} 23
After action of ferment in presence of 1·4 per cent. of sodium carbonate	before sulphuric acid .. after ,, ,, ..	0·084 0·274	} 30
After action of ferment in presence of 1·4 per cent. of citric acid	before sulphuric acid .. after ,, ,, ..	0·014 0·038	} 37

Experiment IV.

(In this instance glycogen was substituted for starch.)

After action of ferment alone	before sulphuric acid .. after ,, ,, ..	0·116 0·238	} 49
After action of ferment in presence of 1·4 per cent. of sodium carbonate	before sulphuric acid .. after ,, ,, ..	0·094 0·384	} 24
After action of ferment in presence of 1·4 per cent. of citric acid	before sulphuric acid .. after ,, ,, ..	0·016 0·028	} 57

The other secretions pertaining to the intestine are those derived respectively from the glands of Brunner and the glands of Lieberkühn.

The secretion of the glands of Brunner, glands special to the duodenum or first portion of the intestine, probably operates in the same way as that of the pancreas in relation to starch digestion; but, from the anatomical situation of the glands, their secretion cannot properly be isolated, and therefore nothing definite has been ascertained with regard to it.

On the other hand, the *succus entericus* or intestinal juice, produced by the glands of the intestine, called glands of Lieberkühn, has been ascertained to have a definite action upon starch, which differs from that of the salivary and pancreatic secretions in possessing a glucose-forming capacity. Under the influence of this secretion, starch is carried through a succession of products possessing an increasing cupric oxide reducing power; but the change does not, as in the case of the salivary and pancreatic ferments, stop at the stage of maltose. If a large amount of the ferment be used, and the action continued for some time, the process of transformation may be found to proceed actually to the production of glucose. Apparently the change is a gradual and continuous one, without any definite halt or interruption; and the point of termination seems to depend upon the amount of ferment and the time during which it is allowed to act. We have here, therefore, under suitable conditions, a glucose-forming ferment to deal with; but the final cupric oxide reducing power presented by a product may stand at any point between maltose and glucose or short of maltose.

The following observations show the results obtained from submitting starch to the action of the ferment existing in the intestine:—

Some starch decoction was mixed with a large quantity of sheep's intestine, and exposed for three hours to a temperature of 48° C. A portion was examined at the end of the three hours, and it was found that a product existed with a cupric oxide reducing equivalent of 50 in relation to glucose at 100. The remainder was allowed to stand for twenty-four hours. The cupric oxide reducing power was then found to have been raised to 85.

Some starch decoction mixed with a large quantity of rabbit's intestine was exposed to a temperature of 48° C. for three hours, and then allowed to stand aside for twenty hours. At the end of the three hours, the product possessed a cupric oxide reducing power a little above that of maltose. At the end of the twenty hours, glucose existed.

Some starch decoction, mixed with a large quantity of rabbits' intestine, was exposed to a temperature of 48° C. for three hours, and then set aside for twenty-four hours. After the three hours' exposure to 48° C., the cupric oxide reducing power was found to be 87. After the subsequent twenty-four hours, it stood at 98.

Some starch decoction was mixed with a watery extract of cat's intestine, and the mixture left for eighteen hours at the ordinary temperature. The cupric oxide reducing power was then found to be 91.

The intestinal mucous membrane is found to retain its transformative power after having been dried, provided the drying has been effected at a temperature insufficient to lead to the destruction of the ferment. Of seven experimental observations, conducted with the dried intestine of the horse and a solution of starch, two yielded results, showing that the sugar produced consisted of glucose. In the other five, the extent of transformation fell more or less short of this, the cupric oxide reducing powers of the several products being represented by the numbers 71, 79, 80, 81, and 86, in relation to that of glucose taken at 100. The temperature employed was 38° C., and the period of exposure to it varied from thirty minutes to four hours. In one of the two cases, in which the product was found to consist of glucose, the mixture had simply been set aside for twenty-four hours at the ordinary temperature.

These observations suffice to show that a ferment exists in the walls of the intestine which has the power of carrying the process of transformation on till glucose is reached. The process passes on gradually through successive stages marked by advancing cupric oxide reducing power. It is evident, however, from the observations to follow, that glucose is not to any large extent produced by the action of the secretions within the alimentary canal, whatever may be the ferment power existing in the intestinal walls themselves.

The contents of the intestine were examined in five rabbits, taken at a period of digestion. The rabbits had been fed in three of the instances on oats and green food, and in the other two on oats moistened with water. An alcoholic extract was made so as to separate the transformed from the untransformed starch. The cupric oxide reducing power of the product obtained was in every case below that of maltose, and where the food had consisted of moistened oats, it was lower than in the others.

Product of Starch Digestion found in the Portal Blood.

We are now brought to the consideration of the question of the precise form in which the changed starch reaches the portal blood, as revealed by observation conducted upon the blood itself. The point touches the practical issue of starch digestion. I have performed a large number of experiments upon the subject. The detailed results will be introduced under the general consideration of the portal blood (pp. 105—106). All that is necessary here to state is that a significant, and often even a very notable, amount of sugar is discoverable in the portal blood as the effect of the ingestion of starchy matter; and that the sugar in its totality has a cupric oxide reducing power varying between that of maltose and that of glucose, thus showing that, whether glucose is present or not, sugar other than glucose must to a greater or less extent exist.

CELLULOSE.

Cellulose is characterised by its resistance to solution by ordinary solvents. Little of a definite nature is known about its solution in the alimentary canal of animals. Probably in the human subject but little cellulose is digested and absorbed. In some of the herbivorous animals, however, notably in the large group of grass and hay feeders, it enters to so large an extent into the food that the capacity for digesting and rendering it of alimentary value may be inferred to exist. This inference is supported by the proof that can be adduced from the vegetable kingdom of cellulose being susceptible of undergoing transformation and solution by the action of certain kinds of ferments. Thus, to quote from Sachs*:—

"The non-nitrogenous reserve material [in the stone of the date] consists of hard cellulose, which is deposited in the endosperm in the form of thickened cell walls, and which constitutes the great mass of the date stone. The embryo, at first very minute, protrudes its root and plumule at the beginning of germination, and only the uppermost portion of the primary seed leaf, which now gradually develops into a cup-shaped absorbing organ, becoming larger and larger, remains within the endosperm. This organ consists of a very delicate parenchyma, and excretes ferments which dissolve the hard endosperm in its immediate neighbourhood. The products of solution are absorbed by the organ and then conveyed into the growing parts of the seedling; until, finally, the whole of the hard date stone is dissolved and its cavity occupied by the developed absorbing organ. The seed of *Phytelephas* (known under the name of vegetable ivory), which is at least a hundred times larger than the date stone, and the

* 'Lectures on the Physiology of Plants.' By Julius von Sachs: Translated by H. Marshall Ward; Clarendon Press, Oxford, 1887.

oudosperm of which consists of much harder cellulose, behaves similarly.

"The action of those fungi which destroy wood and kill trees may be compared directly with the action of such seedlings on their endosperm; the thin mycelial threads of these fungi, as has been shown by Robert Hartig in his magnificent work, penetrate into the alburnum and heart wood of trees, evidently because they excrete ferments at their growing apices, which dissolve the hard cell-walls of the wood."—(p. 344.)

With a suitable ferment present, there is no reason that cellulose should not be susceptible of undergoing digestive solution by transformation into sugar, and of being thereby placed in a position to prove of equal alimentary value to starch or any other carbohydrate. A more or less considerable disappearance of cellulose has been ascertained to take place in the digestive tract, especially of herbivorous animals, which stands in harmony with the hypothesis suggested. But it must be stated that no specific cellulolytic ferment belonging to the alimentary canal has yet been recognised, and the proposition has been advanced that the disappearance of cellulose may be in part accounted for by a transformation of an altogether different kind. Under the influence of putrefactive action cellulose may be broken up, with the evolution of marsh gas as a product, and it has been pointed out that during the detention of the alimentary matter in the very voluminous cæcum of some of the herbivora marsh gas is to a certain extent developed.

Cane Sugar.

Cane sugar, considered as an article of food, is, on account of its ready solubility and diffusibility, in a position widely different from that held by starch, glycogen, and cellulose. As far as its physical properties are concerned, it is, therefore, in a state to be susceptible of absorption from the alimentary canal, without requiring any preparatory process of digestion. It does not, however, follow from this that it becomes absorbed and reaches the portal system without undergoing alteration. This is a point to which attention will be given after certain preliminary considerations have been disposed of concerning the chemical relations of cane sugar, and the methods employed in its estimation.

Cane sugar does not possess any cupric oxide reducing power, but like the other carbohydrates is converted into glucose by boiling with dilute sulphuric acid, the change in the case of cane sugar being known as *inversion*, and its product, as *invert sugar*. Invert sugar is in reality a mixture, in equal quantities, of two forms of glucose—dextrose and lævulose, each of them possessing the cupric oxide reducing power belonging to glucose derived from other sources. A ready means of effecting a quantitative estimation of cane sugar is thus afforded, it being only necessary that the sugar should be subjected to inversion, and the cupric oxide reducing power of the product determined. From this, an expression of the amount of cane sugar is obtainable by calculation. The relation existing between cane sugar and glucose gives as the factor for converting glucose figures into figures representing the corresponding amount of cane sugar, the expression, $\left(\frac{342}{360}\right)$ or 0·95.

Thus :—

Amount of invert sugar (glucose) × 0·95 = Amount of cane sugar.

For conversion of cane sugar into glucose by boiling with sulphuric acid, a 2 per cent. solution of acid is used, as for the other carbohydrates; but the time required for the completion of the process is much shorter than that needed for the other bodies. Full conversion may, in fact, be reckoned to have taken place after boiling for five minutes. I have indeed observed that three minutes have sufficed. The boiling should not be continued beyond the point which will reliably secure complete conversion on account of risk of destruction and the development of colour which would interfere with the subsequent process of titration with the copper test. It may happen that no colour is perceptible at the end of an unnecessarily prolonged boiling, and yet that more or less colour appears when the acid is neutralised.

Cane sugar is, however, convertible into glucose, not only by the mineral acids but also by dilute organic acids, and citric acid is a convenient—indeed, I consider, the most convenient—agent for employment for the purpose. A 2 per cent. strength suffices, and seven minutes are sufficient to allow for the period of boiling. No risk is involved of the production of colour or the occurrence of any destruction. Cane sugar in this respect differs from starch and glycogen and the various products emanating from them. None of these bodies, as I have satisfied myself by direct observation, are affected by boiling with citric acid.

By taking advantage of this property, it becomes practicable to detect and estimate cane sugar in the presence of other bodies (lactose excepted) of the carbohydrate group. The method of procedure consists in dividing the solution to be examined into two equal portions. One portion is titrated at once, and the cupric oxide reducing power belonging to the product ascertained. The second portion is boiled with citric acid (added in proportion to give a 2 per cent. solution), neutralised, and then titrated. An identity between the two results is indicative of an absence of cane sugar. An increase, on the other hand, noticeable after treatment with the acid denotes the presence of cane sugar, and affords a measure of its amount.

It should be mentioned that in all investigations involving the question of the transformation of cane sugar it is necessary to be sure that the specimen employed exerts no cupric oxide reducing action. Specimens often contain more or less glucose as an impurity, and vitiated results may thus be given. The sugar sold in coarse, colourless crystals—that which is known as "white crystal" and is used for sweetening coffee—may be instanced as being of the requisite purity. No cupric oxide reducing power existing at starting, whatever cupric oxide reducing power is subsequently met with may be read as affording a measure of the amount of cane sugar that has undergone transformation.

With these preliminary remarks we may now follow the cane sugar after its ingestion.

Cane sugar taken into the mouth undergoes no change before reaching the stomach. I have submitted the point to actual observation, and my observations, in accordance with what is commonly stated, have furnished no evidence that any action is exerted upon cane sugar by saliva, and I may here further state that the same holds good with regard to the pancreas.

Some confusion exists with regard to what occurs in the stomach in relation to cane sugar. The general belief, expressed in broad terms, is that cane sugar is not acted upon to any extent by the stomach, but that in the intestine it comes into contact with a ferment which transforms it into invert sugar. Neither of these statements must be taken as literally and wholly correct. Observations I have conducted have shown, in the first place, that a ferment exists in the stomach-walls capable of producing a certain small amount of transformative effect; and, in the second, that although the intestine of animals generally possesses an active transformative power, in the ruminant such is not the case. In this class of animals the transformative power is located in certain portions of the stomach.

The subjoined experimental evidence, obtained from various animals, shows in a decided manner that a certain amount of transformative power is, as a general condition, possessed by the stomach as well as by the intestine. It is true, the power in the intestine is very much greater than that in the stomach. This will be seen from the quantitative results to be adduced. In the case of the intestine, the transformative energy is sufficiently strong to produce an effect that is readily perceptible through rough qualitative testing. In the case of the stomach, however, the effect might easily be overlooked if a more delicate method of procedure were not adopted. The following are the results, briefly expressed, of some of the experiments I have conducted.

A solution of cane sugar was introduced into the thoroughly cleansed stomach of a recently killed rabbit. The stomach, ligatured at its two orifices, was then placed in a beaker of water, and, after being exposed for thirty minutes to a temperature of 48° C., was set aside in the laboratory for twenty-four hours. At the end of this time a considerable amount of sugar had diffused through the stomach-walls, and was to be found in the surrounding water. Examination showed that 9 per cent. consisted of glucose. Analysis of the liquid remaining in the stomach showed the presence of glucose to the extent of 24 per cent. of the sugar found.

A coil of intestine was dealt with in a similar way. At the end of the twenty-four hours, examination showed that 40 per cent. of the sugar contained in the diffusate, and 30 per cent. of that present in the contents of the intestine consisted of glucose. In a duplicate experiment with the intestine, 53 per cent. of the diffused, and 45 per cent. of the undiffused, sugar consisted of glucose.

It will be noticed in the above experiments that, on the one hand, a certain amount of sugar had passed through the membranous medium in an untransformed state, and that, on the other, a certain amount of transformed sugar was encountered in the undiffused liquid.

In other experiments, scrapings from the mucous membrane of the stomach and the intestine were placed in contact with a solution of cane sugar, and exposed to conditions favourable to ferment action. With the intestine, in the case of the rabbit, pig, horse, dog, and cat, a marked transformative action was found to be exerted, whilst in the case of the ruminant an absence of effect was observed. With the stomach (the stomach of the ruminant will receive separate consideration later on), there appeared to be evidence of a certain, though comparatively insignificant, amount of transformation, the evidence being more decided in the case of the pig than in the other animals.

The mucous membrane not only possesses transformative power in its fresh state, but also after having been kept. Indeed, the setting in of decomposition appears to lead to the production of an increased transformative energy. The following experimental results may be adduced in support of this statement :—

25 grams of fresh rabbit's intestine were mixed with 0·5 gram of cane sugar dissolved in 50 c.c. of water, and allowed to stand for 15 minutes at 38° C. The product obtained by subsequent treatment with sulphate of soda contained 0·126 gram of transformed cane sugar. In corresponding experiments, made with the same rabbit's intestine which had been previously allowed to remain exposed to the air for periods of 24, 72, and 96 hours, the amounts of glucose found were respectively 0·363, 0·405, and 0·312 gram. Another 25-gram portion of the intestine was allowed to stand for 120 hours, and was then mixed with only half the quantity of sugar previously employed. After proceeding as in the other experiments, examination of the product showed that the whole of the sugar present existed in the form of glucose.

Alcohol precipitates the ferment without destroying its activity. The precipitated material, dried at a suitably moderate temperature, may be kept, and will be found to retain its virtue for an indefinite period. I have recently experimented with a specimen, derived from the mucous membrane of the horse, which has been kept for four years. It yielded the following evidence of activity:—

30 grams of the dried substance in a powdered state were mixed with 5 grams of cane sugar dissolved in 100 c.c. of water, and kept at a temperature of about 38° C. 20 c.c. portions were removed at intervals and examined. At the end of 1½ hours it was found that 17 per cent., at the end of 4½ hours 32 per cent., and at the end of 21 hours (during the latter part of the time the temperature had fallen to 26° C.) 72 per cent. of the sugar present consisted of glucose.

The ruminant animal, as has already been stated, stands in a position different from that of animals generally, in relation to the point under consideration. The intestine possesses only an insignificant amount of transformative energy. The effect produced upon cane sugar by its mucous membrane is about comparable to that which I have described as produced by the stomach of other animals. A compensatory action is exerted elsewhere. The reed, or fourth division, is the portion of the ruminant stomach which represents the ordinary stomach; and, in harmony with this, it is found that only an insignificant power of transforming cane sugar is possessed by its mucous coat. In connexion, however, with the other portions, a more or less marked transformative energy is shown by observation to exist. The following stands in illustration of what I have stated.

Portions of the cleansed paunch, reticulum, manyplies, reed—or true stomach, and intestine of a sheep, were minutely divided, and 10-gram quantities of each taken. After the addition of a little water, they were severally mixed with 5 c.c. of cane sugar solution, containing the equivalent of 0·104 gram of glucose, and exposed for two hours to a temperature of 48° C. Subjoined are the respective amounts of glucose that were found to be present:—

Paunch	0·086 gram.
Reticulum	0·020 ,,
Manyplies	0·068 ,,
Reed, or true stomach	0·016 ,,
Intestine	a trace.

Other observations have been conducted, and though the figures obtained have not uniformly held the same relation, a general agreement with those given above has been noticeable. The paunch and manyplies have given evidence of possessing more transformative energy than the reticulum, and the reticulum than the true stomach.

The experimental evidence adduced in the preceding text has shown that the walls of the stomach possess a certain, though slight, power of inverting cane sugar. The enquiry hitherto has only touched the effects producible by the structural parts of different portions of the alimentary canal. Another point that presents itself for investigation is whether any effect is produced within the stomach by the action of the contents at a period of digestion. I have made a number of experiments upon this point, and will give, as examples, the results obtained in some of them. Evidence, it will be seen, is afforded of the occurrence of a marked amount of transformation.

A dog was fed upon meat, and, during digestion, 30 grams of cane sugar in solution were injected into the stomach through a tube passed down the œsophagus. Twenty minutes later the animal was killed. The liquid contents of the stomach were collected and examined. In a total of 7·500 grams of sugar found to be present, 0·600 gram existed in the transformed state. The contents of the intestine were also examined, and it was found that in a total of 1·470 grams of sugar rather less than half consisted of glucose.

A solution containing 28 grams of cane sugar was given to another dog, which had been fed with meat four hours previously. The animal lapped the liquid out of a vessel. Twenty minutes afterwards it was killed. Of the 6·492 grams of sugar found in the liquid portion of the stomach-contents, 1·081 grams, or about one-sixth, consisted of glucose. In the intestinal contents, taken for examination, the sugar found amounted to 0·645 gram, of which 0·244 gram consisted of glucose. From the figures thus yielded in this and the preceding observation, evidence is presented that cane sugar, to a greater or less extent, reaches the intestine in an untransformed state, and therefore, in part, escapes absorption by the stomach.

In the observations just referred to, the sugar was introduced into the stomach of the living animal. The effect has also been ascertained of bringing cane sugar into contact with the contents of the stomach outside the body. In the observations conducted, the contents of the stomach of rabbits were employed, and exposure for half an hour to a temperature of 48° C. allowed. The sugar was then extracted with alcohol, to eliminate from the product the starchy matter present. From a comparison of the figures representing the amounts of sugar found before and after inversion with those representing the sugar contained in a corresponding amount, similarly dealt with, of the stomach-contents taken alone, evidence was afforded of an extensive transformation of cane sugar having in each instance taken place. In one of the experiments the whole had become transformed.

The effect of what has preceded is to show that when cane sugar is brought into contact with the contents of the stomach, whether within the living animal or not, a marked extent of transformation into glucose occurs. That such a result should ensue is only in accord with what might be expected, seeing that dilute acids have the power of inverting cane sugar, even at moderate temperatures. This I learnt in the course of my experiments, and the fact has not escaped the notice of others. To show the kind of effect produced, I may give the following particulars of two observations on the point:—

1. A liquid containing 3 drops of hydrochloric acid in 100 c.c. of water was placed in contact with 5 grams of cane sugar and kept at 38° C. 20 c.c. portions were removed at the end of five, and of twenty-four, hours. The amounts of invert sugar respectively found in them were 0·157 and 0·392 gram.
2. A solution of citric acid, to represent an organic acid, of 5 per cent. strength was, in the other case, used, and the same method of procedure as in the previous observation adopted. The 20 c.c. portions removed at the end of five and of twenty-four hours respectively contained 0·112 and 0·326 gram of invert sugar.

Product derived from Ingested Cane Sugar found in the Portal Blood.

The next matter for consideration is one which stands as the main point of physiological interest and importance in connexion with the ingestion of cane sugar, viz., the form of sugar which actually reaches the portal blood.

Careful experimenting is required in order to obtain the desired information, and it is necessary, in the first place, to have a knowledge of the condition natural to the portal blood apart from the ingestion of cane sugar. This constitutes the basis for the interpretation of the results obtained after its ingestion. Later on, attention will be given to the state of the portal blood under various conditions (*vide* p. 101, *et seq.*), and it will be advisable to defer till then entering into detailed considerations of its condition after the ingestion of cane sugar. It will here be sufficient to state that an increased amount of sugar is met with, and that glucose is the form of sugar present. In my own experiments no indication has been afforded of the passage of cane sugar, as such, into the portal blood, but the statement is made that after the ingestion of large amounts of cane sugar traces may be found both in the blood and the urine. I learnt from observations (*vide* p. 116) conducted many years ago, that, under the circumstances named, glucose in notable amount became discoverable in the urine, but it did not occur to me to examine and ascertain whether cane sugar was also present.

LACTOSE.

Lactose is a readily soluble and diffusible body, and, by virtue of these properties, does not require to undergo change within the alimentary canal to be rendered fit for absorption. I have subjected lactose to the action of the coats of the stomach and intestine of the rabbit, cat, and sheep, and have not obtained evidence of a decided nature of any transformation being brought about. It is stated, however, that, under the influence of inverting ferments, in the same manner as by the action of acids, lactose is convertible into galactose and dextrose, and that the *succus entericus* has the power of producing this effect. Whether or not the transformation occurs physiologically, proof, it may be assumed, is afforded of the possibility of its occurrence within the system by the fact that in the diabetic ingested lactose appears as glucose in the urine.

Lactose, as is well known, readily undergoes the lactic acid fermentation in presence of certain micro-organisms. Apparently a certain amount of the lactose reaching the alimentary canal becomes transformed in this way.

GLUCOSE.

Glucose stands as the end product inside the carbohydrate series of the action of ferments, &c., on the various members of the group. No change can therefore be looked for under exposure to the influence of the digestive secretions, and none is required to be exerted, seeing that, by virtue of its high solubility and diffusibility, it exists in a fit state for passage into the circulation by absorption. After ingestion, it can be traced into the portal blood. Like lactose, though less readily, glucose is susceptible of undergoing the lactic acid fermentation, and this change, under certain circumstances, may, to some extent, occur in the alimentary canal.

PORTAL BLOOD IN RELATION TO INGESTED CARBOHYDRATES.

I have traced the ingested carbohydrates on to the point at which they stand as soluble diffusible bodies within the alimentary canal, in close proximity to the vessels distributed upon its inner surface, and thus in a favourable position for absorption into the circulation to occur. I will now pass to the examination of the portal blood, and see what evidence is adducible of their passage into it.

In entering upon this enquiry, it is necessary in the first place to know the constitution of the portal blood, in relation to sugar, apart from the influence of ingested carbohydrate matter. The blood of the portal system contains a certain amount of sugar derived from that existing as a constituent of the blood of the general circulation. As I shall show in detail later on (p. 161), the blood of the general circulation contains a standard or definite amount of sugar, which, under natural and ordinary conditions, may be stated to range from about 0·6 to 1·0, or a little over 1·0, per 1000, and which presents no evidence of any essential variation in the different parts of the system spoken of. In considering, therefore, the question of the effect produced upon the portal blood by the absorption of ingested carbohydrate matter, the existence of this initial amount of sugar, common to the blood as a whole, must be taken into account.

We start, then, with blood containing from about 0·6 to about 1·0 per 1000 of sugar, due to that belonging naturally to the general circulation.

With this preliminary observation, I will proceed to give results showing the condition of the portal blood found:—

 a. At a period of fasting.
 b. After the ingestion of animal food.
 c. After the ingestion of food rich in carbohydrate matter.

Some years ago I conducted an extensive series of experiments upon the point under consideration, and the results showed a marked increase of sugar after the ingestion of carbohydrate material. Looking, however, at the importance of the matter from the point of view of the conclusions that may be based upon it, I have recently undertaken, in the Research Laboratories of the Colleges of Physicians and Surgeons, a number of additional experiments, which have been conducted with the advantage of the analytical experience gained from previous work. The results of these experiments, it may be stated, exhibit a general conformity with those formerly obtained. I will here give them in their entirety.

In each case the animal was suddenly killed by pithing. Instantly afterwards, an incision was made through the abdominal parietes, and the portal vein ligatured. The blood flowing to the liver was thus caused to accumulate below the ligature, whilst at the same time any backward flow from the organ was prevented. Blood was then collected direct from the vessel. In some of the instances, it will be noticed, two portions were collected and analysed. In these cases the flow was for a few moments stopped, by compression of the vessel, before the collection of the second portion.

It will be observed in the analytical results to follow that the kind of sugar found in the portal blood is one possessing a cupric oxide reducing power standing ordinarily below, and sometimes considerably below, that of glucose. As will be seen from what will be stated later on, a point of difference here exists between the sugar of the portal blood and that present in the general circulation, which, under natural circumstances, is found to have the cupric oxide reducing character of glucose.

Portal Blood at a Period of Fasting.

			Sugar per 1000, expressed as glucose.	Relation of the cupric oxide reducing power of the sugar present to that of glucose at 100.
I.	Dog, nearly 48 hours after last food:			
	Portal blood	{ before sulphuric acid { after ,, ,,	0·580 0·643	} 90
II.	Dog, nearly 48 hours after last food:			
	Portal blood	{ before sulphuric acid { after ,, ,,	0·666 0·710	} 94
III.	Rabbit, fully 24 hours after last food:			
	Portal blood	{ before sulphuric acid { after ,, ,,	1·583 1·733	} 91
IV.	Rabbit, fully 24 hours after last food:			
	Portal blood	{ before sulphuric acid { after ,, ,,	2·454 2·499	} 98
V.	Rabbit, fully 24 hours after last food:			
	Portal blood	{ before sulphuric acid { after ,, ,,	2·244 2·244	} 100

In the case of the two dogs (Experiments I and II), it will be observed that the blood was collected after a fast of nearly 48 hours. It may be concluded that influence from ingestion would be here absolutely excluded. The amount of sugar found stood within the natural range given for the contents of the general circulation. The prolonged fasting consequently, on its part, did not cause any sensible diminution.

In the case of the rabbits (Experiments III, IV, and V), a different condition existed. Notwithstanding that fully twenty-four hours had elapsed since the last food was taken, the blood gave evidence of containing sugar derived from ingestion. In this animal, it is to be remarked, the stomach is known to contain a considerable quantity of material even after a much more prolonged period of fasting, and doubtless the condition found is thus to be accounted for. Seeing that the animal ordinarily will not be without food for twenty-four hours, the circumstances, it will be perceived, are here such as to lead to the portal blood containing as a constant condition more sugar than that of the general circulation.

Portal Blood, after Animal Food.

		Sugar per 1000, expressed as glucose.	Relation of the cupric oxide reducing power of the sugar present to that of glucose at 100.
I.	Dog, fed for 3 days on lean beef; killed 4 hours after last food; abundant chyme in intestine:		
	Portal blood. 1st portion collected { before sulphuric acid	1·117	} 83
	after ,, ,,	1·340	
	Portal blood. 2nd portion collected { before sulphuric acid	1·335	} 82
	after ,, ,,	1·631	
II.	Dog, fed for 3 days on lean beef; killed 4 hours after last food; no chyme visible in intestine:		
	Portal blood. 1st portion collected { before sulphuric acid	0·893	} 82
	after ,, ,,	1·087	
	Portal blood. 2nd portion collected { before sulphuric acid	0·743	} 73
	after ,, ,,	1·023	
III.	Dog, fed for 2 days on lean beef; killed 6¼ hours after last food; apparently in full digestion:		
	Portal blood .... { before sulphuric acid	0·633	} 93
	after ,, ,,	0·683	
IV.	Dog, fed for 2 days on lean beef; killed 5¼ hours after last food:		
	Portal blood .... { before sulphuric acid	0·593	} 90
	after ,, ,,	0·660	
V.	Dog, fed for 2 days on lean beef; killed 5½ hours after last food:		
	Portal blood .... { before sulphuric acid	0·633	} 90
	after ,, ,,	0·707	
VI.	Dog, fed on lean beef; killed 5 hours after last food:		
	Portal blood. 1st portion collected { before sulphuric acid	0·750	} 96
	after ,, ,,	0·776	
	Portal blood. 2nd portion collected { before sulphuric acid	0·693	} 85
	after ,, ,,	0·817	
VII.	Dog, fed on lean beef; killed 5¼ hours after last food:		
	Portal blood .... { before sulphuric acid	0·720	} 100
	after ,, ,,	0·720	
VIII.	Cat, fed on meat; killed 5 hours after last food:		
	Portal blood .... { before sulphuric acid	0·851	} 77
	after ,, ,,	1·139	
IX.	Cat, fed on meat; killed 5 hours after last food:		
	Portal blood .... { before sulphuric acid	0·675	} 87
	after ,, ,,	0·779	

Portal Blood after the Ingestion of Starchy Food.

		Sugar per 1000, expressed as glucose.	Relation of the cupric oxide reducing power of the sugar present to that of glucose at 100.
I.	Dog, shortly after having been fed with a little meat; 22 grams of starch in 200 c.c. of water injected into stomach through an œsophageal tube; killed 30 minutes afterwards:		
	Portal blood .... { before sulphuric acid after „ „	1·210 1·267	} 96
II.	Dog, fed on bread and milk; killed 3¼ hours afterwards:		
	Portal blood. 1st { before sulphuric acid portion collected after „ „	1·183 1·477	} 80
	Portal blood. 2nd { before sulphuric acid portion collected after „ „	1·763 2·500	} 70
III.	Dog, fed on bread and milk; fed badly; killed 4 hours afterwards:		
	Portal blood .... { before sulphuric acid after „ „	1·153 1·270	} 91
IV.	Dog, fed on bread and broth; killed 2 hours afterwards:		
	Portal blood .... { before sulphuric acid after „ „	0·977 1·100	} 89
V.	Dog, fed on bread and broth; killed about 1 hour afterwards:		
	Portal blood .... { before sulphuric acid after „ „	0·968 1·277	} 76
VI.	Dog, fed on bread and broth; killed 2 hours afterwards:		
	Portal blood .... { before sulphuric acid after „ „	1·723 2·187	} 78
VII.	Dog, fed on bread, milk, and meat; killed 2¼ hours afterwards:		
	Portal blood .... { before sulphuric acid after „ „	1·403 1·930	} 73
VIII.	Rabbit, fed on moistened oats; killed 2 hours afterwards:		
	Portal blood .... { before sulphuric acid after „ „	2·587 3·234	} 80
IX.	Rabbit, fed on moistened oats and green food; killed 3 hours afterwards:		
	Portal blood .... { before sulphuric acid after „ „	2·346 2·878	} 82
X.	Rabbit, fed on moistened oats; killed 4 hours afterwards:		
	Portal blood .... { before sulphuric acid after „ „	1·630 2·085	} 78

Portal Blood after Ingestion of Starchy Food (continued)

			Sugar per 1000, expressed as glucose.	Relation of the cupric oxide reducing power of the sugar present to that of glucose at 100.
XI.	Rabbit, fed on moistened oats; killed 4 hours afterwards:			
	Portal blood	{ before sulphuric acid { after ,, ,,	2·109 2·903	} 73
XII.	Rabbit, fed on moistened oats; killed 4 hours afterwards:			
	Portal blood	{ before sulphuric acid { after ,, ,,	1·431 1·816	} 79
XIII.	Rabbit, fed on moistened oats; killed 4½ to 5 hours afterwards:			
	Portal blood	{ before sulphuric acid { after ,, ,,	3·665 4·102	} 89
XIV.	Rabbit, fed on moistened oats; killed 4½ to 5 hours afterwards:			
	Portal blood	{ before sulphuric acid { after ,, ,,	4·590 5·330	} 86
XV.	Rabbit, fed on moistened oats; killed 4½ to 5 hours afterwards:			
	Portal blood	{ before sulphuric acid { after ,, ,,	4·070 4·690	} 87
XVI.	Rabbit, fed on moistened oats; killed 5 to 5½ hours afterwards:			
	Portal blood	{ before sulphuric acid { after ,, ,,	1·461 1·657	} 88
XVII.	Rabbit, fed on moistened oats; killed 5 to 5½ hours afterwards:			
	Portal blood	{ before sulphuric acid { after ,, ,,	1·442 1·615	} 89
XVIII.	Rabbit, fed on moistened oats; killed 5 to 5½ hours afterwards:			
	Portal blood	{ before sulphuric acid { after ,, ,,	1·413 1·567	} 90
XIX.	Rabbit, fed on moistened oats; killed 5 to 5½ hours afterwards:			
	Portal blood	{ before sulphuric acid { after ,, ,,	1·693 1·698	} 99
XX.	Rabbit, fed on moistened oats; killed shortly after being fed:			
	Portal blood	{ before sulphuric acid { after ,, ,,	1·407 1·707	} 82

PORTAL BLOOD AFTER MALTOSE.

Osazone from the sugar of the portal blood of rabbits at a period of digestion
Magnified 400 diameters.

Portal Blood after the Ingestion of Maltose.

	Sugar per 1000, expressed as glucose.	Relation of the cupric oxide reducing power of the sugar present to that of glucose at 100.

I. Dog, 24 hours after last food; 32 grams of maltose in 85 c.c. of water injected into stomach through an œsophageal tube; killed 25 minutes afterwards:

Portal blood .... { before sulphuric acid 1·600 } 72
                  { after   ,,   2·200

Portal Blood after the Ingestion of Cane Sugar.

To ascertain if cane sugar passes as such into the portal blood, it is necessary to resort in the process of analysis to the employment of citric acid instead of sulphuric acid. Alcoholic extraction is undertaken in the usual way, but at the end of the evaporation sulphate of soda must not be used, as the presence of this agent, it is found, more or less interferes with the inverting action of organic acids upon cane sugar. A less satisfactory product is given for titration, but this is unavoidable. After filtration, preferably through glass wool, has been performed, the liquid is brought to a given bulk and divided into two portions. One portion is titrated at once, whilst the other is boiled for seven minutes with 2 per cent. citric acid, neutralised, and then titrated. Any increase revealed by the second titration is representative of cane sugar.

		Sugar per 1000, expressed as glucose.
I.	Dog, shortly after having been fed with a little meat; 33 grams of cane sugar in 200 c.c. of water injected into stomach through an œsophageal tube; killed 30 minutes afterwards:	
	Portal blood. 1st portion collected { before citric acid....	1·333
	{ after ,, ...	1·333
	Portal blood. 2nd portion collected { before citric acid....	1·520
	{ after ,, ...	1·522
II.	Dog, not fed since previous day; 70 grams of cane sugar in 210 c.c. of water injected into stomach; killed 30 minutes afterwards:	
	Portal blood .... { before citric acid....	2·113
	{ after ,, ...	2·113
III.	Dog, not fed since previous day; 70 grams of cane sugar in 210 c.c. of water injected into stomach; killed 30 minutes afterwards:	
	Portal blood .... { before citric acid....	1·844
	{ after ,, ...	1·600

It will be noticed in the first two observations that the figures yielded before and after citric acid stand in strikingly close accord. In the third a disparity exists which can only arise from error of analysis, seeing that the figures after citric acid are lower than those before. The inference from these results is that carbohydrate derived from the ingested cane sugar reached the portal blood as glucose and not in the state of cane sugar.

THE LIVER IN RELATION TO THE SUGAR DERIVED FROM INGESTED CARBOHYDRATES.

From the results which have preceded, evidence is afforded that ingested carbohydrate matter passes in the form of sugar from the alimentary tract into the blood of the portal system. By the portal system the sugar is conveyed to the liver, and the next point for consideration is the question of what now becomes of it. Does it pass through the organ and reach the general circulation, or is it stopped and disposed of in some other manner?

The examination of this question may be approached from two sides : from the one, by comparing the condition of the blood taken before and after passing through the liver; and, from the other, by looking to the condition of the liver itself, in relation to the influence exerted by the ingestion of carbohydrate food. Let us see what light is thrown on the matter by appeal to experimental investigation, looking first at the information derivable from an examination of the blood.

1. *Evidence afforded by the Blood of Sugar being stopped by the Liver.*

Seeing, as has been shown in the preceding pages, that the blood of the portal system after the ingestion of carbohydrate matter contains more sugar than that of other parts of the circulation, we have a fact to deal with which points in two directions. It implies not only entry of sugar by absorption through the capillary radicles of the portal system in the walls of the alimentary canal, but also exit by abstraction on reaching the terminal capillaries in the liver.

It is the latter point, namely, the abstraction of sugar from the portal blood by the liver, that we have now to consider, and its consideration requires that attention should be given to the condition of the blood flowing from, as compared with that flowing to, the organ.

The portal vein, after entering the liver, terminates in capillaries, from which the hepatic vein takes origin and passes to empty itself into the inferior cava. The blood, therefore, to be compared is that contained, on the one hand, in the portal, and, on the other, in the hepatic vein. The collection of blood in a suitable state for examination from the former is easily made. It is also within the range of possibility to make the collection from the latter. It can be done during life by passing a properly constructed catheter down through the right jugular, the superior cava, the right auricle of the heart, and the inferior cava, to the seat of entrance of the hepatic veins. The operation, however, like other operations on the living animal attended with a disturbance of the tranquil and natural state, conduces to the production of an altered condition of the blood in relation to sugar, and cannot be relied upon to yield the desired information, which must therefore be sought for in some other way.

How speedily an alteration in the state of the blood, attended with the presence of an abnormal amount of sugar, may take place through the influence of altered conditions connected with the liver, I illustrated at the outset of my researches, now upwards of thirty years ago. From the readiness with which the blood flowing from the liver is thus thrown into an altered state, it is necessary that close attention should be given, and proper precautions observed, to escape being misled; otherwise, sugar may be met with more or less greatly in excess of what is ordinarily present, and the inference thence be erroneously drawn that its discharge from the liver takes place in a manner that does not naturally occur.

In dealing with the problem before us, it is, then, a matter of primary importance that the blood taken for examination should be in a state representative of that naturally belonging to life. To obtain such a specimen directly from the hepatic veins is, I consider, a procedure attended with so much liability to the introduction of error as to be unsuited for employment. With the blood of the right side of the heart, however, in which that flowing from the liver is present, mixed with that derived from the systemic veins, no such difficulty exists. This is easily obtained in a state representative of that naturally belonging to life, and, through its examination, information may be indirectly obtained regarding the constitution of the hepatic blood of which it in part consists. Later on (*vide* p. 164), I shall enter into details about the operative procedure to be adopted. Suffice it here to state that I consider the best method is to make the collection instantly after the sudden destruction of life. With the observance of quickness throughout, the blood may thus be obtained before time has been given for any change of condition to occur.

The amount of sugar found in the blood collected from the heart in the manner just described ranges, as I have previously stated, from about 0·6 to 1·0, or a little over 1·0, per 1000—the amount agreeing with that met with in the blood of the systemic veins as it is circulating under natural conditions. If an influx of sugar took place through the hepatic veins, the effect should be visible in the cardiac, on being compared with the general systemic venous, blood; but observation does not show that any recognisable difference exists between the two.

I may mention that it has sometimes happened, in my experiments for collecting blood for examination from the right side of the heart, that, in the haste of manipulation, the scissors have been plunged into the chest in too much of a downward direction, and that the diaphragm and contiguous portions of the liver adjoining the inferior cava have been incised, without the heart being touched. Under these circumstances, the blood which escaped was derived, in considerable amount, directly from the divided hepatic veins, but the analytical results subsequently obtained showed no perceptible difference from those yielded by the blood procured from the heart.

The position, then, of the matter actually before us for consideration stands thus:—The portal blood, after the ingestion of carbohydrate matter, contains sugar amounting to from about 1·5 to 2, 3, 4, or even more per 1000, whilst the blood on the other side of the liver, under similar circumstances, does not afford evidence of containing more than from about 0·6 to 1·0, or a little over 1·0, per 1000. The conclusion to be drawn is that the surplus amount of sugar found in the portal blood is abstracted from it during its passage through the vessels of the liver.

It will be noticed that what I have asserted is in direct opposition to the tenets of the glycogenic theory propounded by Bernard. The discrepancy, as shown elsewhere in this work, is attributable to a want of recognition of the fact that the amount of sugar found in the blood on the cardiac side of the liver, in the absence of the precautions requiring to be observed in procuring it for examination, is largely in excess of what is natural.

The office performed by the liver, of abstracting sugar from the portal blood, has the effect of maintaining the contents of the general circulation in a state of uniformity. If the sugar derived from ingested carbohydrate matter were allowed to pass through the liver and reach the general circulation, a disturbance of this uniformity would be constantly occurring. Such a state of things, indeed, is precisely what exists in diabetes. Whilst, in health, the blood of the general circulation is shielded from variations due to the influence of ingested carbohydrate matter, and the urine remains impregnated (*vide* p. 186) with only an insignificant amount of sugar, corresponding with the small amount existing naturally in the blood: in diabetes, sugar reaches the general circulation, and thence the urine, in proportion as carbohydrate matter enters the portal system from ingestion. The blood of the portal system, in the natural order of things, is variable in character, from the influence exerted by ingestion; and should it happen that such variability of character is allowed to be transmitted to the blood on the other side of the liver, the result occurring is a proportionate escape of sugar with the urine—in other words, the production of the condition belonging to diabetes.

2. *Evidence afforded through the Liver itself of the Stoppage by it of Sugar derived from Ingestion.*

Having dealt with the question under consideration from the point of view of the comparative condition of the blood on the two sides of the liver, we have next to look to the state of the liver itself to see if evidence is obtainable of the detention of the carbohydrate matter conveyed to it by the blood from the alimentary tract.

In entering upon the examination of this question, I am carried back to investigations conducted almost at the commencement of my association with physiological research, some thirty-five years ago. The announcement had been made by Claude Bernard, in 1848, that the liver possessed a sugar-forming function. Bernard had set before himself the task of ascertaining what became, within the system, of the sugar derived from ingested carbohydrate matter. In the prosecution of his inquiry, he conducted observations upon animals which had been kept on food of a saccharine nature, and was led to conclude that he had traced the sugar through the liver into the general circulation as far as the right side of the heart. In order to prove that the sugar here met with had been derived from the food, he performed a counterpart experiment upon an animal which had been kept upon meat only, and, contrary to expectation, still found sugar in the contents of the circulatory system, in the same manner as before. Here arose the starting point for his glycogenic theory. At first, as a part of the theory, he regarded the production of sugar as due to a vital process of the nature of secretion, looking upon albuminoid matter as contributing to its formation. Subsequently, however, he discovered the body from which the sugar actually takes origin, and, from the physiological position which he regarded it as holding, he gave it the name of glycogen. In the course of his investigations, he found that this material underwent transformation into sugar, not only after removal of the liver from the animal, but even after the passage of a stream of water through the vessels, to thoroughly wash out the blood. He further extracted and isolated the newly discovered body, and found that it consisted of a carbohydrate. The announcement of its isolation was made in the year 1857.

Up to this time, our knowledge had not been brought into a more advanced position than this. But the new line of research which was opened out excited a great amount of interest in the scientific world, and drew various workers into the field of enquiry. I stood as one of these, and, at the same time, as one who had been an eye-witness of Bernard's experimental work; and, in the prosecution of the researches I conducted, I found that the principle which had been recognised as the source of sugar in the liver was itself derived from ingested carbohydrate matter. This discovery was made in the year 1858, and the particulars relating to the subject were given in a communication published in the 'Phil. Trans.' for 1860 (p. 579). I will here introduce an epitomised representation of the points of evidence brought forward in the communication referred to.

I will direct attention first to the observations conducted upon dogs, and in these observations it will be understood that all the animals were in good condition, and in no way specially selected. In each case, the dog was fed for some days upon the particular food the influence of which upon the liver it was desired to examine, and was then killed by pithing, after being weighed. The liver was at once removed and weighed, without the gall bladder and after the blood had been allowed to drain off. In most instances a determination was also made of the amount of glycogen present. The process employed for the determination of glycogen had not, it should be stated, then attained the satisfactory position of that now at our disposal, in which the estimation is effected by converting into glucose and determining the amount of this principle by means of the ammoniated cupric test. At that time the plan adopted was simply to boil the liver with potash and then to precipitate the glycogen by pouring into spirit. This precipitate was weighed and reckoned as glycogen, and, although not consisting purely of glycogen, it gave a fair representation of the relative amount existing in different livers.

To serve as a basis of comparison, eleven dogs were killed after being restricted to a diet of purely animal food. In these instances the average relation of liver-weight to body-weight was as 1 to 30. The actual figures ranged from 1 to 33, as the lowest, to 1 to 21, as the highest, the latter instance standing alone in giving so high a relative weight of liver. A determination was made of the percentage of crude glycogen precipitate yielded by seven of the livers, and the figures obtained ranged from 4·88 to 10·95, averaging 7·19.

To obtain information as to the effect upon the liver of a diet of vegetable food with its preponderance of carbohydrate material, five dogs were fed for several days on barley-meal and potatoes, or, when this was refused, on bread and potatoes. The average weight of the liver was here found to be one-fifteenth of the body-weight, the figures expressing the actual ratios standing thus :—

1 to 14·5
1 „ 14·5
1 „ 21·0
1 „ 10·5
1 „ 22·5

No analyses were made of the livers of the first two of these dogs, but the quantity of glycogen, as evidenced by the strongly milky character of the decoctions yielded, was unusually large. In fact, it was whilst examining these livers in relation to the question of sugar that I was first led to notice the effect of vegetable food, to which I am now referring. The liver of the third dog was not examined until an hour and a half after death. The crude glycogen precipitate then yielded amounted to 9·87 per cent. In the fourth instance the liver gave 25·30 per cent. of crude glycogen precipitate, and in the fifth 16·50 per cent. The average percentage of crude glycogen precipitate for the three livers was therefore 17·23, a result standing in marked contrast to the amount obtained after animal food.

In another set of observations sugar was given in conjunction with animal food. The sugar employed consisted of cane sugar—the brown or moist sugar in common use. Four dogs were subjected to this regimen.

1. A nearly full-grown dog was kept for eight days on a diet consisting of a bundle of tripe, with at first $\frac{1}{8}$ lb. and afterwards $\frac{1}{4}$ lb. of sugar per diem. The liver-weight was found to stand to the body-weight as 1 to 13·5, and the liver yielded 12·8 per cent. of crude glycogen precipitate.

2. A young dog was fed for nine days on a bundle of tripe and $\frac{1}{4}$ lb. of sugar daily. The ratio between liver- and body-weight stood as 1 to 14·5, the liver yielding 17·55 per cent. of crude glycogen precipitate.

3. A full-grown dog was kept for eight days on a daily allowance of a bundle of tripe and $\frac{1}{4}$ lb. of sugar. The weight of the liver stood to the weight of the body as 1 to 26. The percentage of crude glycogen precipitate yielded was 12·33.

4. A nearly full-grown dog was fed for five days on a diet similar to that last mentioned. Here the liver-weight was one-fourteenth of the body-weight, and the percentage of crude glycogen precipitate 15·37.

The four examples here given present, as the average of liver-weight to body-weight, the ratio 1 to 15, and an average of 14·5 as the percentage of crude glycogen precipitate yielded by the liver: results pretty closely corresponding in character with those observed after a diet of ordinary vegetable food.

In both of these last sets of results it is noticeable that the liver is markedly increased in size and weight as compared with what is found to obtain after a diet of animal food, and this increase stands in harmony with the large proportion of crude glycogen precipitate yielded.

With reference to the observations on the effect of an admixture of sugar with animal food, it may be incidentally mentioned that the liver was found to present an appearance strikingly different from what is usual. After purely animal food it is comparatively dark coloured, and of so firm and fleshy a consistence as to require considerable pressure to break it down between the fingers. After the administration of sugar, however, the liver was found to be pale coloured, with a tinge of pink, and soft enough in consistence to be readily broken down by a very slight pressure. It had the appearance of being swollen and flabby.

Another point that is worthy of being mentioned is that in three out of the four instances where sugar was administered, the urine collected from the bladder after death was found to contain sugar in the form of glucose, although cane sugar had been administered.

Proceeding now to the experiments on the rabbit, we find that the evidence yielded is such as to show in a simple and conclusive manner that ingested carbohydrates lead to the formation of glycogen in the liver.

1. A couple of full-grown rabbits, as closely as possible resembling each other in weight and condition, were taken for experiment. One was kept fasting, whilst the other was fed daily for three days, through a flexible tube passed down the œsophagus to the stomach, with 1 oz of starch and $\frac{3}{4}$ oz. of grape sugar, made into a semi-fluid mass with water. On the fourth day both animals were killed. The weight of the fasting animal was 3 lbs. 1 oz., that of the other 3 lbs. 4 ozs. The liver of the fasting animal weighed $1\frac{2}{5}$ ozs., that of the other $2\frac{4}{5}$ ozs., or exactly twice as much.

The liver of the rabbit fed on starch and grape sugar was rich in glycogen, yielding 15·4 per cent. of crude precipitate, whilst that of the other rabbit only yielded 1·3 per cent.

In another experiment two half-grown rabbits, likewise as nearly as possible resembling each other, were made the subjects of a comparative observation. One was kept fasting, whilst the other was fed daily for three days in the same manner as before, with 1 oz. of starch and 1 oz. of cane sugar instead of grape sugar as in the former observation. On the fourth day both animals were killed, and the examination conducted. The weight of the fasting animal was 1 lb. 14 ozs., and that of the other 1 lb. $14\frac{3}{4}$ oz. The liver of the fasting rabbit weighed 1 oz., and that of the other $2\frac{1}{4}$ ozs., or considerably more than twice as much. The liver of the rabbit fed on starch and cane sugar yielded 16·9 per cent. of crude glycogen precipitate, whilst the liver of the other yielded only 1·4 per cent.

As in the case of the dogs after the administration of sugar with animal food, the livers of the rabbits fed on starch and sugar were of a very pale colour, and so soft as to be readily broken down by slight pressure between the fingers. This, as I have already remarked, is the condition presented by the liver when rich in glycogen. In an instance of unusual richness that happened to fall under my observation, the liver was, indeed, so soft as to be almost pulpy, scarcely holding together when taken up by a pair of forceps.

The result of experience since these observations were made is to confirm the conclusion that was drawn from them—namely, that the ingestion of carbohydrate matter leads to the presence of an increased amount of glycogen in the liver. This, indeed, may now be regarded as an accepted physiological fact, and it is turned to account in the practice which is resorted to of feeding an animal largely with carbohydrate matter previously to its being killed, when it is desired to obtain glycogen in quantity from the liver.

Looking, then, to the information derivable alike from the blood and from the liver, we find evidence is afforded that the sugar emanating from ingested carbohydrate matter is carried to the liver by the portal blood and there checked in its progress, instead of being allowed to pass on and reach the general circulation. It is through the agency of the hepatic cells that this effect is produced, and within them a concurrent accumulation of glycogen has been observed by micro-chemical examination to take place. Thus the diffusible carbohydrate which has reached the portal blood becomes abstracted and transmuted by the hepatic cells into the non-diffusible form of it—glycogen. The process of transmutation that here occurs constitutes an instance of the carrying down of carbohydrate matter to a lower state of hydration, an operation of the reverse nature of that which is noticed to be brought about by agencies acting apart from the influence of living matter. The transformations effected in the laboratory by ferment action and chemical agents are, as is known, transformations attended with increased hydration.

Animal Food as a Source of Glycogen.

The issue of what has preceded is that ingested carbohydrate matter can be traced from the alimentary canal through the portal system to the liver, where it becomes stopped in its progress, and transformed from the state of sugar into that of glycogen, with the result that the amount of the latter present in the organ fluctuates according to the amount of carbohydrate matter ingested.

It is known, however, that glycogen is discoverable in the liver apart from the ingestion of saccharine and starchy articles of food. Under a diet of purely animal food, glycogen, to a certain extent, is found to exist, and its origin has been referred to a breaking up of nitrogenous matter within the liver itself. Another explanation, however, which carries us back to what takes place antecedent to absorption and stands in harmony with the general train of events occurring in connexion with digestion, is susceptible of being given.

In the first place, it may be said that animal food contains a certain amount of sugar. Probably the statement is within the limits of truth that no living substance exists which does not contain proteid, carbohydrate, fatty, and mineral matter, and, in accord with this proposition, flesh and other animal substances consumed as food can be shown by analysis to contain free carbohydrate under the form of sugar, and, to some extent also, of glycogen. Sugar from these sources will, as a result of the ordinary operations of alimentation, reach the portal vein and thence the liver.

Ordinary beef tea may be selected to illustrate the point in question, and subjoined is a photo-engraving of a micro-photograph fo the osazone derived from the sugar present.

Osazone from the sugar present in ordinary beef tea.
Magnified 400 diameters.

In the next place, besides the free carbohydrate existing as a constituent of animal substances, there exists, as I have shown in what I have said when speaking of the glucoside constitution of proteid matter, carbohydrate in a locked up state. Not only is this carbohydrate susceptible of being liberated by the cleaving agency of acids and alkalis, but likewise by that of ferment action, it having, as a matter of fact, been experimentally shown (p. 50) to be set free by ordinary pepsin digestion. For the recognition, however, of the liberated carbohydrate material, the copper test is not satisfactorily available, on account of the masking effect of the peptone concurrently produced. With the phenylhydrazine test, it happens that the peptone does not similarly constitute a source of difficulty, osazone crystals being obtainable in its presence. Subjoined is a representation of osazone crystals derived from a purchased sample of peptonised meat.*

* "Fluid meat" prepared by Savory and Moore.

Osazone from the sugar present in peptonised meat.
Magnified 400 diameters.

In animal food, then, we have a certain amount of free carbohydrate under the forms of both sugar and glycogen, and likewise a source of carbohydrate in the cleavage action of digestion upon proteid matter. As with vegetable food, in fact, carbohydrate is supplied, though to a less extent, for alimentation; and, in harmony with this, the results given by examination of the portal blood may be taken, I consider, as affording evidence of the occurrence of carbohydrate absorption. If reference be made to the collection of results that have been given of the examination of portal blood after animal food, it is noticeable that, whilst in several instances no positive evidence, it is true, is perceptible of augmentation of sugar from absorption, yet in others there are indications pointing to its existence. Moreover, in some of the instances in which first and second portions of blood were collected, the latter showed the presence of a larger quantity of sugar, suggesting that with the longer detention in the capillary vessels a greater absorption occurred. It must be remembered that with the ligature of the portal vein the flow of blood through the liver is checked, and that the contents of the circulation are thereby prevented from being influenced in the manner that ordinarily occurs as the effect of sugar production after death.

There is evidence, as will elsewhere in this work be shown, suggesting that a certain amount of carbohydrate is applied to the formation of proteid and fat by the agency of the protoplasmic chemical power possessed by the cells of the intestinal villi; and it follows that, in proportion as the carbohydrate absorbed from the contents of the alimentary canal is thus diverted, less will be left for passage in the free state into the portal blood. In this way it may happen that the evidence of absorption afforded by the portal blood is less certain and pronounced than it otherwise would be.

From the foregoing considerations it is seen that a source for the glycogen of the liver exists in animal alimentation in every way comparable to that existing in connexion with vegetable alimentation. The only difference is one of degree in the amount of carbohydrate-supply, and this harmonises with the difference in the respective amounts of glycogen found to be present in the liver under the two kinds of alimentation.

Authorities speak of glycogen taking origin from proteid matter within the liver by a proteolytic action exerted by its cells. It has been in this way considered that the presence of glycogen is to be accounted for under subsistence upon animal food. I have shown, however, that another explanation can be given, without looking to the liver for the performance of a proteolytic office. The source of the glycogen is sufficiently to be accounted for by the occurrence of the same kind of action as that which is in operation after the ingestion of vegetable carbohydrate food. It is the province of ferment action to break up and hydrate, and of protoplasmic action to synthesise and de-hydrate. The proposition, therefore, which I have advanced stands in accord with the natural order of events, whilst to assume that the protoplasmic liver-cells perform the work of ferment action does not. That they should de-hydrate or transmute sugar into glycogen is, on the other hand, consistent with what can be shown to be accomplished by protoplasmic matter elsewhere.

General Relations of Glycogen as a Constituent of the Liver.

I have hitherto been speaking of the glycogen of the liver in relation to its origin from ingested carbohydrate and proteid substances. I have now to direct attention to its general relations as a constituent of the liver.

Glycogen is recognisable in the liver-cells by micro-chemical examination, and is represented as existing in a hyaline state, together with fat globules and proteid granules, in the meshes of the protoplasmic network of the cell. As the result of treatment with iodine, the glycogen is brought into view by the red-brown staining that occurs. It may be dissolved out by water, leaving the network of the cell-protoplasm intact. In spite of its ready solubility, its extraction by water from the liver-substance is not, however, easily to be effected, owing, it may be assumed, to its non-diffusible nature.

It is known that even a diffusible body like glucose requires sedulous treatment with water for its complete extraction from the tissues. With glycogen, the difficulty is very much greater, and even after the liver has been brought into a state of comparatively fine subdivision by pounding in a mortar, and has been subjected to successive boilings with water until no signs of further extraction are afforded, a considerable quantity remains unremoved. If, afterwards, the residue be set aside in a little water until the following day, the liquid will be found to have become again milky in appearance from the further extraction that has ensued. The same occurrence may be noticeable even with several successive daily repetitions of the treatment, and finally the residue on examination may still be found to contain a considerable quantity of glycogen. I drew attention to this circumstance in a communication presented to the Royal Society in 1881; and from the detailed observations therein narrated I will select one for the purpose of illustration.

A weighed quantity of the liver of a recently killed dog was reduced to a pulp in a mortar and thoroughly extracted with alcohol for the removal of its sugar. The coagulated residue from the alcoholic washings was then repeatedly extracted with boiling water till the liquid came away clear, instead of lactescent, from the presence of glycogen, as at first. The residue was kept in a moist state till the following day, when the process of washing was repeated. On the third and fourth days the process was again repeated until in each case a clear liquid was obtained. The glycogen in each day's washings was estimated by conversion into glucose and recourse to titration with the ammoniated cupric test. Finally, the residue from the successive washings was boiled with potash and poured into spirit for the collection and estimation of the glycogen remaining. The figures yielded stood as follows :—

	Per 1000 of liver.
Glycogen extracted on the 1st day	13·833
,, ,, ,, 2nd ,,	3·879
,, ,, ,, 3rd ,,	2·961
,, ,, ,, 4th ,,	2·817
,, remaining in the residue	35·145

It will thus be seen that the repeated extractions on the four successive days removed only something under half of the total amount of glycogen contained in the liver. The above results, together with others from similarly conducted observations, show how imperfectly ordinary extraction with water removes glycogen from the liver-substance, and account for the small quantity, compared with what might be expected, that is not unfrequently obtained when the process of collection by aqueous extraction is adopted.

The extent to which glycogen is extracted by water is largely dependent upon the degree to which the minuteness of subdivision of the liver-substance has been carried, as the following experiment, taken in conjunction with the one that has been given, contributes to show.

A portion of the liver of a freshly killed rabbit was plunged into boiling water for the purpose of checking the loss of glycogen by *post-mortem* change. It was then pounded in a mortar, and afterwards still further reduced to a finely-divided state by being forcibly squeezed through muslin. A weighed quantity was now boiled with successive portions of water until no appearance of lactescence was observed. The treatment with boiling water was repeated on the second day, and again on the third. After the third day's extraction, the residue was once more thoroughly pounded, and again boiled with water. The extract yielded was strongly lactescent, notwithstanding that the liquid at the end of the previous washing was perfectly clear. After this further extraction, the residue was boiled with potash, and the glycogen precipitated by alcohol and quantitatively determined in the usual way. The following are the results that were obtained :—

					Per 1000 of liver.
Glycogen extracted on the 1st day				....	36·629
,,	,,	,,	2nd	,, ...	7·156
,,	,,	,,	3rd	,, ...	1·904
,,	,,	,,	3rd	,, after further pounding	6·107
,,	remaining in the residue			....	6·540

The presence of acetic acid appears to promote extraction. A second portion of the liver made use of in the preceding experiment was extracted with successive portions of water containing 0·5 per cent. of acetic acid. As will be seen from the subjoined figures, more glycogen was extracted than where water alone was employed.

					Per 1000 of liver.
Glycogen extracted on the 1st day				....	43·533
,,	,,	,,	2nd	,, ...	8·085
,,	,,	,,	3rd	,, ...	2·421
,,	,,	,,	3rd	,, after further pounding	2·097
,,	remaining in the residue			....	1·620

It is a point worthy of note that, notwithstanding the difficulty with which extraction is effected by water at 100° C. (212° F.), the whole of the glycogen is readily and speedily extracted when the operation is performed under pressure in a digester worked at a temperature of about 140° C. (284° F.). Half an hour is found to suffice for the purpose, as will be seen by reference to the subjoined results obtained in two experiments in which a comparison was made through the application of the potash process. Two rabbits were taken, and the livers removed and treated with alcohol for the extraction of the sugar. In each case duplicate portions were submitted to comparative examination with the use of the high pressure digester (autoclave) and the application of the ordinary potash process. The following were the figures yielded :—

	Glycogen per 1000 of liver.	
	Obs. 1.	Obs. 2.
After half an hour's aqueous extraction at 140° C.	93·015	68·040
After boiling with 10 per cent. potash for half an hour	91·314	64·242

Considerations regarding the Amount of Glycogen in the Liver.

The amount of glycogen encountered in the liver presents a wide range of variation. It is governed not only by influences in operation during life, but also by influences coming into operation after death. The former have been in part already referred to, and will be further referred to later on. As regards the latter, the position that we have to deal with is this: the liver, like other parts of the system, contains a certain amount of carbohydrate. The amount, however, in the liver is, under ordinary circumstances, much larger than that found elsewhere. In form, it mainly consists of glycogen, but, precisely as elsewhere, there is present a certain small proportion of sugar. The carbohydrate susceptible of extraction by alcohol, which is reckoned as *sugar*, here presents a cupric oxide reducing capacity below that of glucose, and sometimes even considerably below that of maltose. In the latter case, something of a dextrin-like nature must obviously be present.

The glycogen is placed under conditions which lead to its speedy transformation into dextrin and sugar after death, and, unless precautionary measures are adopted to check this *post-mortem* transformation, a loss of glycogen in proportion to the dextrin and sugar produced will occur. In order, therefore, to obtain a representation of the amount of glycogen existing at the time of death, it is necessary either to instantly check *post-mortem* change, as by rapidly plunging the liver of a suddenly killed animal into a freezing mixture, or else to make allowance for the glycogen which has disappeared by conversion into sugar subsequently to death.

Where the liver is taken after having been dealt with in a manner to prevent the *post-mortem* production of sugar, the result given by analysis represents the amount of glycogen corresponding with that existing during life, associated with the sugar also naturally belonging to life which may be said ordinarily to amount to from 2 to 3 per 1000. Where, on the other hand, the organ is taken for examination in an ordinary way, a less amount of glycogen than there should be is given, whilst the sugar accompanying it may be found, according to the extent to which *post-mortem* change has occurred, to amount to 8, 10, or 12, or from this to 15, 20, or even, it may be, 25 per 1000.) Deduction of the amount of sugar naturally belonging to the liver during life, which I have put at from 2 to 3 per 1000, from that found when *post-mortem* change has not been checked will give the amount produced after death, which stands equivalent to the amount of the transformed glycogen. From these data it is possible, in the case of the liver taken in an ordinary manner for examination, to estimate, with a close approximation to the truth, the amount of glycogen existing at the moment of death. No matter, then, whether the liver be taken with special precautions to prevent loss of glycogen from *post-mortem* transformation or not, an estimation of its amount may be made with sufficient accuracy to admit of physiological conclusions being drawn.

It must be borne in mind that it follows from what was said in the section of this work on "the glucoside constitution of proteid matter" that the analytical results taken as expressive of glycogen include the cleavage carbohydrate liberated by the potash process adopted. The amount of this material is, however, too small to affect the validity of the conclusions deducible from the results, where glycogen to any significant extent is present. It stands otherwise, it must be said, where only insignificant figures are yielded. These, in reality, may be either wholly or partially due to cleavage carbohydrate, and thus are not to be read as necessarily implying that any glycogen is actually present.

The material before me, serving as the basis of the remarks about to be made upon the amount of glycogen found in the liver under various conditions, consists of upwards of 200 quantitative estimations, extracted from my laboratory books. These I do not put forward as exhaustive of the matter, but they suffice to afford indications of a general nature.

In the process of analysis, the carbohydrate is estimated in the form of glucose, but since the body to be represented is one of the composition $C_6H_{10}O_5$ instead of $C_6H_{12}O_6$, it follows that the result must be multiplied by 0·9 in order to obtain an expression of the true amount of glycogen present. With the figures that follow this has been done.

A survey of the results shows the existence of a wide range of variation in the glycogen figures that have been obtained. The lower limit may be placed as low as 1 or 2 per 1000, or possibly, even less. The amounts commonly found range from 4 or 5 to 20 or 30 or even 40 per 1000. Such figures as 60, 70, and 80 per 1000 are not unfrequently met with, and I have come across amounts as large, for example, as 120, 122, and 126 per 1000. These last quantities were found in dogs which had been specially supplied with food rich in carbohydrates. The animals yielding respectively the figures 120 and 122 per 1000 had been kept for some days upon a diet of sugar, bread, meat, and milk; and the one which gave 126 per 1000, for ten days upon bread and meat. The amount standing next was 114 per 1000. It was yielded by a rabbit which had been taken without any special feeding.

Of the various conditions affecting the amount of glycogen in the liver the most potent is food. Reference has already been made (p. 113, *et seq.*) to the influence of carbohydrate food in leading to an increased production and accumulation of glycogen, and nothing further need here be said with regard to it.

The liver of the young animal, whether taken in the foetal stage or subsequently, appears to contain more glycogen than is found in later life. Amongst my laboratory records I find an instance in which a dog had been kept for some days upon a diet of meat and sugar, for the purpose of ascertaining the condition of the organs in relation to glycogen. When killed, it was found to be in pup, and the organs of the foetus, as well as those of the parent, were examined.

The liver of the parent, taken in an ordinary way after death, yielded 27 per 1000 of glycogen, accompanied with 22 per 1000 of sugar. The livers of the two foetal pups examined showed respectively the presence of 59 and 63 per 1000 of glycogen, with 8 and 7 per 1000 of sugar.

I have also records of the condition existing in two cats with sucking kittens. In one case the liver of the parent yielded 31 per 1000 of glycogen, and the livers of the kittens, taken together, 42 per 1000. No determination of the sugar was made in this instance. In the other, the liver of the parent was found to contain 21 per 1000 of glycogen, accompanied with 21 per 1000 of sugar, whilst the liver of one kitten yielded 32 per 1000 of glycogen with 17 per 1000 of sugar, and that of another 45 per 1000 of glycogen with 10 per 1000 of sugar.

The condition of the liver in different animals in relation to glycogen may next be referred to. I have before me the results of a number of observations which have been from time to time conducted. These I will consider in groups arranged under the heads of the several kinds of animals examined. It will be understood that the animals at the time of being taken were under ordinary conditions, and that in the great majority of the instances the examination was conducted in an ordinary way without the observance of special precautions for checking *post-mortem* change. Under these circumstances, allowance has to be made for the loss of glycogen occurring as a *post-mortem* event, and in order that this may be done the figures representing the sugar will be inserted after those expressive of the glycogen found. In a certain number of the instances the liver was taken after having been plunged instantly after death into a freezing mixture, whereby the loss of glycogen by *post-mortem* change was prevented.

Dog.—In many of the observations before me, referring to this animal, there had been a special dieting with carbohydrate food. These observations do not properly fall within what is intended to be here represented, and they will therefore be excluded from consideration, with the remark that the figures expressive of glycogen were largely in excess of those yielded by the liver under the ordinary diet of animal food. In nine instances of dogs fed only upon meat the livers gave glycogen figures as follows:—3·3 per 1000 (with 17 per 1000 of sugar), 4·2 per 1000 (with 14 per 1000 of sugar), 5·5 per 1000 (with 5 per 1000 of sugar), 8·7 per 1000 (with 19 per 1000 of sugar), 8·9 per 1000 (with 13 per 1000 of sugar), 12·2 per 1000 (with 16 per 1000 of sugar), 18·0 per 1000 (with 6 per 1000 of sugar), 26·4 per 1000 (with 17 per 1000 of sugar), and 39·6 per 1000 (with 19 per 1000 of sugar).

Cat.—I have a record of a large number of observations upon this animal. In some instances there had been special dieting with bread and milk along with the meat, and with reference to these special instances I consider I need not say more than that most of the results failed to show an increase of glycogen to anything like the same extent as that noticed in many of the livers of dogs fed in a similar way. In one case, however, the glycogen figures stood at 65·3 per 1000 (with 5 per 1000 of sugar), and in another at 58·6 per 1000 (with 19 per 1000 of sugar). In most of the instances they were quite low. I cannot undertake to reconcile the discrepancy which has been referred to, but the suggestion may be thrown out that possibly the cats may not have properly taken the bread and milk provided. In the twenty-four instances where it is recorded that the animals had been kept upon meat only, the glycogen figures varied from 1·8 per 1000 (with 16 per 1000 of sugar) to 35·3 per 1000 (with 21 per 1000 of sugar). Only in a few cases did the glycogen figures, taken irrespectively of the sugar, stand higher than 20 per 1000, and in more than half they stood under 10 per 1000.

Rabbit.—Twenty-nine observations, including fourteen in which the liver had been frozen instantly after death. The highest glycogen figures stood at 114·1 per 1000 (with 2·4 per 1000 of sugar). Next to these were 93·1 per 1000 (with 2·6 per 1000 of sugar), 81·8 per 1000 (with 1·1 per 1000 of sugar), and 78·5 per 1000 (with 2·1 per 1000 of sugar). In five other cases the glycogen figures, taken alone, were in excess of 50 per 1000. The lowest figures obtained were 5·2 per 1000 (with 4·6 per 1000 of sugar), 5·8 per 1000 (with 15·2 per 1000 of sugar), and 7·3 per 1000 (with 2·8 per 1000 of sugar). The others ranged between these extremes.

Bullock.—Two observations. Glycogen figures yielded—21·0 per 1000 (with 24 per 1000 of sugar), and 5·9 per 1000 (with 15 per 1000 of sugar).

Calf.—Two observations. Glycogen figures yielded—22·0 per 1000 (with 21 per 1000 of sugar), and 13·0 per 1000 (with 14 per 1000 of sugar).

Sheep.—Eight observations. In two, where the animals had been specially fed, under my directions, upon barley-meal for three days previous to being slaughtered, the glycogen figures obtained stood only at 1·3 per 1000 (with 10 per 1000 of sugar), and 3·4 per

1000 (with 15 per 1000 of sugar). In four, where the animals had been specially fed upon maize for two days before death, the livers yielded glycogen figures as follows:—8·2 per 1000 (with 17 per 1000 of sugar), 11·1 per 1000 (with 15 per 1000 of sugar), 11·5 per 1000 (with 25 per 1000 of sugar), and 21·4 per 1000 (with 25 per 1000 of sugar). In the other two, in which I have no record of the nature of the food, and in which the liver was kept for some considerable time before being examined, the glycogen figures stood respectively at 6·0 per 1000 (with 23 per 1000 of sugar), and 0·9 per 1000 (with 30 per 1000 of sugar).

Pig.—Two observations. Glycogen figures yielded—6·0 per 1000 (with 13 per 1000 of sugar), and 8·2 per 1000 (with 7 per 1000 of sugar).

Horse.—Six observations. In two, nothing is recorded about the condition of the animal or the nature of the food. The glycogen figures yielded stood at 30·0 per 1000 (with 16 per 1000 of sugar), and 11·5 per 1000 (sugar not determined). In two others, the remark is entered that the animals were in poor condition, and had been fed on hay. The figures representative of glycogen stood at 12·8 per 1000 (with 12 per 1000 of sugar), and 11·0 per 1000 (with 3 per 1000 of sugar). The remaining two animals had been specially fed upon maize for two days before being killed. One is described as being old and in bad condition, and here the glycogen figures obtained were 29·5 per 1000 (with 15 per 1000 of sugar); the other was in fair condition, and in this instance the glycogen figures yielded were 73·1 per 1000 (with 20 per 1000 of sugar).

The conditions ordinarily attaching to the butcher's slaughter-house are not such as to conduce to the liver being found rich in glycogen. There is, for instance, the more or less prolonged interval between the withdrawal from the accustomed conditions of the stall or pasture and the arrival at the market, with the fatigue incidental to the journey. Then, there is the indefinite period of detention in the slaughterhouse. And, finally, the liver, after death, is likely to remain for a longer time under conditions permitting of loss of glycogen by *post-mortem* transformation than is allowed to be the case in the laboratory.

In the case of the horse the amount of glycogen is seen to range higher than in that of the animals from the butcher's slaughterhouse.

It appears from the statements of other observers, that horseflesh is coincidently rich in glycogen. Apart, however, from any specific difference that may exist, there is the point for consideration that the horse has usually been less unnaturally circumstanced previous to death than the other animals in question, on account of its not having had to undergo the sufferance involved in passing through the market. Although, it is true, many horses are killed on account of old age, or, it may be, some internal disease, yet it frequently happens that the animal is merely disabled from work by a local affection of the legs, which need not derange its general health.

Domestic Fowl.—One observation. Glycogen figures yielded—2·4 per 1000 (with 3·8 per 1000 of sugar).

Grouse.—One observation. Glycogen figures yielded—1·7 per 1000 (with 3·7 per 1000 of sugar).

Tortoise.—Two observations. Glycogen figures yielded—21·7 per 1000 (with 2·2 per 1000 of sugar), and 86·7 per 1000 (with 1·4 per 1000 of sugar).

Frog.—One observation. Glycogen figures yielded—43·3 per 1000 (with 1·7 per 1000 of sugar).

Cod-fish.—Two observations. Glycogen figures yielded—1·8 per 1000 (with 1·5 per 1000 of sugar), and 4·0 per 1000 (with 2·6 per 1000 of sugar).

Mackerel.—One observation. Glycogen figures yielded—2·2 per 1000 (with 3·0 per 1000 of sugar).

Salmon.—One observation. Glycogen figures yielded—0·4 per 1000 (with 3·6 per 1000 of sugar).

Lobster.—Three observations. Glycogen figures yielded—3·0 per 1000 (with 2·0 per 1000 of sugar), 5·0 per 1000 (with 3·4 per 1000 of sugar), and 2·0 per 1000 (with 2·5 per 1000 of sugar).

Crab.—Two observations. Glycogen figures yielded—2·7 per 1000 (with 8·1 per 1000 of sugar), and 8·3 per 1000 (with 4·6 per 1000 of sugar).

Oyster.—Five observations. Glycogen figures yielded—27·5 per 1000 (with 2·3 per 1000 of sugar), 9·9 per 1000 (with 2·2 per 1000 of sugar), 27·2 per 1000 (with 1·5 per 1000 of sugar), 41·4 per 1000 (with 2·8 per 1000 of sugar), and 13·5 per 1000 (with 4·5 per 1000 of sugar).

Mussel (Mytilus edulis).—One observation. Glycogen figures yielded—17·4 per 1000 (with 1·9 per 1000 of sugar).

THE LIVER IN RELATION TO SUGAR.

Having dealt with the carbohydrate matter which exists in the liver under the form of a material insoluble in alcohol and devoid of cupric oxide reducing power—that is, a material comprehended under the term *glycogen*—I have next to treat of the carbohydrate matter soluble in alcohol, possessed of cupric oxide reducing power, and comprehended under the general term *sugar*. This latter, whilst forming a constituent of the liver, is likewise, as will further on be shown, found to be present universally throughout the tissues and organs of the animal system, as far as they have been examined.

If the carbohydrate now being considered is merely subjected to the ordinary processes of sugar detection and sugar determination, the information is not yielded that is necessary for supplying a knowledge of its character, or form. It would under the circumstances presumably be taken as consisting of glucose, whilst in reality the form of sugar present might be different. If, however, it is subjected to the process of examination to which I have resorted in my investigations, a process which embraces the determination of the cupric oxide reducing power before and after boiling with sulphuric acid, data are afforded for supplying more extended information, and it may be found that, instead of glucose, we are dealing with a product of lower cupric oxide reducing power. The cupric oxide reducing power, indeed, that in different instances may be met with is found to vary from a low one upwards, in the manner that would occur from the presence of mixtures of dextrins and sugars. Although, as is obvious, the carbohydrate material present cannot, when possessing a cupric oxide reducing power below that of maltose, be entirely composed, strictly speaking, of sugars, yet for physiological purposes it may, for the sake of convenience and brevity, be broadly comprehended under the term.

With these preliminary remarks, I will proceed to the consideration of the relation of the liver to sugar taken in the sense just defined.

Examination of the liver reveals the existence of a certain amount of sugar. The amount found varies according to the conditions under which the examination is made. If the organ be taken in such a manner as to represent as closely as possible the condition belonging to life, the sugar met with is insignificant in amount, whilst, if time be allowed to elapse between the death of the animal and the removal of the liver, sugar is found to be largely present.

In 1860, through a communication presented to the Royal Society and published in the 'Transactions' for the following year, I brought under notice the difference existing in the condition of the liver taken, on the one hand, at the moment of death, and, on the other, a short time afterwards.

Previously, from the time of the promulgation of Bernard's glycogenic theory, the strongly saccharine condition of the liver met with under ordinary examination after death had been looked upon as representing the condition normally existing during life. I had, a short time before, made the discovery that the blood of the right side of the heart was not in the saccharine condition during life that had been previously inferred from the examination of blood removed in an ordinary manner after death. Although I had thus recognised the difference in the *ante-mortem* and ordinary *post-mortem* states of the blood, yet, in common with others, I had not at the time any idea that the saccharine state of the liver revealed under the method of examination then adopted was likewise due to a *post-mortem* change. Not being able to understand how the difference in the two states of the blood was to be accounted for, I was driven to look to the liver in search of an explanation. I first tried the effect of injecting blood at different pressures through the organ after death to imitate different states of blood pressure in the vessels, and ascertain if this influenced the escape of sugar. On failing to obtain any satisfactory information from these experiments, the idea occurred to me to subject the liver to examination to see whether there might not be the same difference between the *ante-mortem* and *post-mortem* states as had been noticed in the blood. I did not start with the expectation that anything would issue from submitting the question to the test of observation, but I nevertheless considered the matter worthy of trial, and hence resolved to endeavour to obtain from the liver a representation approaching as closely as possible to that of the actual living state, by supplying conditions at the instant of death to prevent the occurrence of *post-mortem* change in relation to sugar.

It was known that the liver contained a substance susceptible of rapid transformation into sugar by ferment action, and that a ferment capable of effecting the transformation was present. The object, therefore, to be attained was to deprive the ferment of activity at the instant of death.

In my first experiments I sought to effect this by injecting a strong solution of potash into the liver through the portal vein. As the result, traces only of sugar were afterwards found in the organ.

Subsequently, I resorted to a method of experimenting in which the desired object of checking *post-mortem* ferment action was attained through the simple physical agency of alteration of temperature. By sufficiently elevating the temperature, the ferment is coagulated and its activity destroyed, whilst, by sufficiently lowering it, ferment action is suspended.

Plunging a portion of liver, excised as quickly as possible after death, into boiling water leads to the destruction of the ferment, and thus prevents any subsequent production of sugar. Obviously, however, the effect of increase of heat up to a point short of that necessary for the destruction of the ferment will be to promote transformation, and it must be borne in mind that with a thick mass of liver an appreciable amount of time will be required for the effectual penetration of the heat to the deeper portion, in which the opportunity will thus be given for the occurrence of a certain amount of change. In spite of this circumstance, I found, on performing the experiment, that the condition presented by the liver in relation to sugar stood in striking contrast to that which had been previously supposed to belong to it.

By immersion, on the other hand, in a freezing mixture, as of ice and salt, ferment action is checked, without the ferment being destroyed. Contrary to what occurs with the application of heat, the effect of exposure to cold operates continuously in the desired direction, the energy of the ferment being diminished in proportion as the temperature is lowered, until it is brought to a state of inactivity. If the liver to be examined is only of moderate thickness, as is the case with that of the rabbit, it will become frozen throughout with sufficient rapidity to effect what is wanted; but if a thick mass of material has to be dealt with, such as occurs with an excised portion of the liver of a dog, a few incisions should be made across it to promote the more rapid penetration of the freezing influence.

Upon the grounds just stated, it will be intelligible that, of the two methods of checking *post-mortem* change, that by freezing may be regarded as preferable; but it must not be forgotten in its application that the ferment has not undergone destruction, and that, therefore, the capacity for sugar production still exists, and will come into play should the opportunity be given, in the succeeding processes of preparation for examination, for it to do so. Observation, in fact, shows that a piece of frozen liver which has been allowed to thaw and is afterwards set aside at an ordinary temperature contains in some hours' time about as much sugar as a piece of the liver which has not been frozen.

The ice and salt mixture employed should be prepared about half an hour before it is wanted for use, in order that the occurrence of a certain amount of liquefaction may place it in a favourable position for acting rapidly upon the immersed piece of liver. A minute or two will then be found to suffice for bringing the specimens into a hard, frozen state.

At the now distant period when this method of experimenting was initiated, the subsequent treatment of the frozen liver consisted simply in the preparation of a decoction and the examination of this for sugar by boiling in a test-tube with the copper solution. For the preparation of the decoction, thin slices were pared off the frozen liver and pounded to a pulp in a cooled mortar.

A capsule with a little water in it was next placed over a flame, and after the water was brought to a brisk state of ebullition, the liver pulp was introduced into it, a little at a time, in a manner to secure the instantaneous destruction of the ferment. After a few minutes, the liquid was strained off, and a portion of it boiled in a test-tube with Fehling's solution, when it was found that, instead of the immediate and copious formation of yellow or red precipitate encountered in examinations of the liver conducted without the observance of the precautions referred to, the contents of the test-tube remained blue, and only after standing exhibited a slight subsidence of red oxide particles.

Such was the method of experimenting by which I showed the

fallacy of taking the results of ordinarily conducted examinations of the liver in a *post-mortem* state as representing the condition existing during life. The strongly saccharine condition of the liver, which had been erroneously assumed to represent the physiological state, formed one of the two main points of consideration that led to the evolution of the glycogenic theory propounded by Bernard; the other main point being the character taken to belong naturally to the blood issuing from the liver, which I shall, later on, show to have been equally founded on error.

Since the time when my original experiments were conducted, considerable advance has been made in analytical procedure, and we are now placed in a position to express ourselves in language other than that founded upon the results of mere qualitative testing.

Through extraction with alcohol, subjection of the product to the inverting action of sulphuric acid, and the employment of the ammoniated cupric test, we can now not only express in definite numerical terms the precise amount of sugar existing, but also indicate the nature of the sugar that is present.

In the application of quantitative determination to the frozen liver, the process to be adopted is as follows: a weighed portion of the pulp obtained by pounding in a cooled mortar is treated with alcohol in a manner to secure the thorough extraction of the sugar. Once thoroughly permeated by alcohol, the liver substance is no longer in a condition to undergo change. Through the influence exerted by the spirit, the ferment is thrown into a state of suspended activity. After dividing the product of alcoholic extraction into two portions, one is titrated at once with the ammoniated cupric test, whilst the other is titrated after having been previously subjected to the inverting influence of sulphuric acid. From the two sets of figures obtained information is given concerning the character of the sugar present.

The amount of sugar encountered in the experiments that I have conducted, in the manner described, upon the dog, cat, and rabbit, has been found to stand at about 2 to 3 per 1000, but I have amongst my records instances in which it was found to stand as low as about 1 per 1000. If the removal and freezing of the liver have not been expeditiously performed, higher figures must be looked for.

The amount of sugar, on the other hand, ordinarily encountered in a liver removed a few minutes after death, and not subjected to special treatment for the arrest of ferment action, may be said to stand at somewhere about 12 to 15 per 1000; more will be found as time goes on, and, after the lapse of 18 to 24 hours, the quantity ordinarily met with amounts to from about 20 to 30 or 35 per 1000. Naturally, the amount will vary, according to the circumstances, as regards temperature, &c., existing. A necessary factor for the occurrence of sugar production is, of course, the presence of glycogen, but I think it may be said that when glycogen is present in very large quantity the production proceeds less actively and less extensively than when it exists in moderate amount. It seems as if the ferment power became lessened by the large accumulation existing. Apparently, a more active production occurs where the liver is allowed to remain full of blood after the destruction of life than where the blood has been permitted speedily to escape from the vessels. The amount of sugar found after the lapse of a few minutes from death as compared with that found after the lapse of several hours shows that the production proceeds with much greater activity at first than it does later on.

As representative examples of the state existing at the moment of death, I may give the results of four experiments performed in the Research Laboratories of the Royal Colleges of Physicians and Surgeons whilst the manuscript for these pages was in course of preparation. Four healthy good-sized rabbits were taken, without selection, from a hutch. After death by pithing, the livers were as expeditiously as possible removed and plunged into a freezing mixture, and subsequently dealt with in the manner that has been described.

Amount and Nature of Sugar found in the Liver promptly removed and frozen.

		Sugar per 1000, expressed as glucose.	Cupric oxide reducing power of the sugar present in relation to that of glucose at 100.
Rabbit A.	before sulphuric acid.. after ,, ,, ..	1·601 2·142	} 75
Rabbit B.	before ,, ,, .. after ,, ,, ..	2·344 3·248	} 72
Rabbit C.	before ,, ,, .. after ,, ,, ..	1·946 2·828	} 69
Rabbit D.	before ,, ,, .. after ,, ,, ..	1·664 2·400	} 69

To give completeness to the experiments, quantitative determinations of the glycogen were made, and the amounts found stood as follows: in rabbit A, 76; B, 59; C, 7; and D, 114 per 1000.

I have given examples representative of the state of the liver in relation to sugar at the instant of death. I will now give further examples, also consisting of recently performed experiments, in illustration of the difference found to exist in the state of the organ at the instant of death and at subsequent periods. Two rabbits were killed by pithing, rapidly opened, and a portion of the liver of each excised and plunged into a freezing mixture. A few minutes later, the other portion was removed, part of it taken at once for examination, and the remainder set aside until the following day, when it was also taken for analysis.

Amount and Nature of Sugar in the Liver at the Moment of Death and at subsequent periods.

Rabbit E.			Sugar per 1000, expressed as glucose.	Cupric oxide reducing power of the sugar present in relation to that of glucose at 100.
Frozen liver	before sulphuric acid..		2·000	} 88
	after ,,	,, ··	2·260	
Liver, a few minutes after death	before ,,	,, ··	12·940	} glucose
	after ,,	,, ··	12·940	
Liver, left till the following day	before ,,	,, ··	34·340	} 93
	after ,,	,, ··	36·820	
Rabbit F.				
Frozen liver	before ,,	,, ··	0·980	} 92
	after ,,	,, ··	1·060	
Liver, a few minutes after death	before ,,	,, ··	11·550	} 95
	after ,,	,, ··	12·130	
Liver, left till the following day	before ,,	,, ··	33·280	} glucose
	after ,,	,, ··	32·530	

As regards the nature of the sugar, it is to be stated that, whilst in the liver taken ordinarily after death the sugar present is usually in the form of glucose, that met with in the frozen liver is usually, on the other hand, found to possess a cupric oxide reducing power of a lower degree—a cupric oxide reducing power, indeed, often approaching or even, it may be, standing below that of maltose.

The following are photo-engravings from micro-photographs of osazone crystals yielded respectively by the liver frozen at the instant of death and the liver taken ordinarily after death.

Osazone crystals from the sugar of a rabbit's liver frozen instantly after death. Magnified 400 diameters.

140   THE LIVER IN RELATION TO SUGAR.

Acicular glucosazone crystals from the sugar of a dog's liver taken in an ordinary way after death. Magnified 400 diameters.

Whilst it is necessary, in dealing with the liver of the warm-blooded animal, to observe the precautions which I have particularised in order to obtain a correct representation of the state belonging to life, in the case of that of the cold-blooded animal the conditions are such as not to require that similar expedients should be had recourse to. With the low body temperature existing, there is not the same rapid *post-mortem* production of sugar occurring, and, if no unnecessary delay is permitted, an ordinarily conducted examination suffices. Subjoined are examples of the condition met with in animals of this kind, and I will allow the figures to speak for themselves.

Amount and Nature of Sugar in Livers of Cold-blooded Animals.

		Sugar per 1000, expressed as glucose.	Cupric oxide reducing power of the sugar present in relation to that of glucose at 100.
Reptile.			
Tortoise.	before sulphuric acid	1·250	} 57
	after ,, ,, ...	2·170	
Tortoise.	before ,, ,, ...	0·852	} 59
	after ,, ,, ...	1·437	
Amphibian.			
Frog ...	before ,, ,, ...	1·316	} 76
	after ,, ,, ...	1·724	

The above three observations were conducted during an October month. The temperature was mild for the time of year. The animals were killed, and the livers afterwards in an ordinary way removed and submitted to analysis. In the case of the frog, fifteen livers were taken for the analysis conducted. A large amount of glycogen in each of the observations was found to exist.

The frog may be made use of to exemplify the modifying influence exerted by the body-temperature existing at the time of death upon the results obtained from an ordinarily-conducted examination of the liver. The above figures were derived from frogs exposed to ordinary conditions in an atmosphere, as stated, of medium temperature. A different result would have been obtained if, previously to death, the body-temperature had been raised, as is susceptible of being done, by placing the animal in an artificially-heated atmosphere. Under these circumstances, the liver is found to stand in a position corresponding with that of the warm-blooded animal.

I have referred to this matter in my previous writings, and in these writings I cited an experiment in which I exposed frogs for a couple of hours to the influence of a temperature of 32° C. (90° F.). An examination of the livers, conducted in the ordinary way, revealed, with the employment of Fehling's solution, the presence of a notable amount of sugar, whilst the livers of a duplicate set of frogs which had not been similarly exposed to the elevated temperature failed to give evidence of a similar nature.

The phenomenon to which I have been alluding did not escape the notice of Bernard. In the 'Comptes Rendus' of the Academy of Sciences, March, 1857, he stated that by lowering the temperature of a batch of frogs sugar may be made to disappear from the liver, and that on afterwards exposing them to warmth it is found to reappear. He adds that it is possible to produce this singular alternation of appearance and disappearance of sugar several times, without any food being given, solely by acting upon the circulation through the medium of temperature. Whilst thus noticing the fact, Bernard missed its true interpretation, which is connected not in reality with life activity, but with activity coming into operation after death and contingent upon the temperature that may happen to exist.

Fish.					Sugar per 1000, expressed as glucose.	Cupric oxide reducing power of the sugar present in relation to that of glucose at 100.
Cod	{	before sulphuric acid		......	1·220	} 81
	{	after	,,	,, ...	1·500	
Cod	{	before	,,	,, ...	2·560	} 98
	{	after	,,	,, ...	2·620	
Mackerel	{	before	,,	,, ...	2·500	} 83
	{	after	,,	,, ...	3·000	
Salmon .	{	before	,,	,, ...	2·230	} 61
	{	after	,,	,, ...	3·630	

The above livers were obtained from the fishmonger, and were therefore derived from animals which had been for some time dead. As often happens with livers obtained from the various animals employed for consumption as food the total amount of carbohydrate matter was found to be small. Unless glycogen is present in quantity at the time of death, there is not the source for a large amount of sugar, and, in the case of the fish in question, the amount of glycogen found associated with the sugar was small, ranging below 4 per 1000.

LIVER OF INVERTEBRATE ANIMALS.

Crustacean.		Sugar per 1000, expressed as glucose.	Cupric oxide reducing power of the sugar present in relation to that of glucose at 100.
Lobster	{ before sulphuric acid { after ,,	1·730 1·980	} 87
Lobster	{ before ,, ,, { after ,, ,,	2·150 3·400	} 63
Crab	{ before ,, ,, { after ,, ,,	7·250 8·100	} 88
Crab	{ before ,, ,, { after ,, ,,	6·600 8·280	} 79

It is noticeable that the amount of sugar found in the case of the crabs was much larger than in that of the lobsters, and it is to be remarked that, whilst the lobsters when taken for examination were in a lively and active state, the crabs, on the other hand, showed scarcely any signs of life. The amount of glycogen present was in each instance small.

Mollusk.—From the group of the Mollusca I have taken the oyster and mussel for examination. The liver does not here exist in a separate form, as in the higher animals. It is interwoven with other structures, and these were included in what was taken for analysis. So large a proportion of the animal, however, consists of liver that the figures obtained may be regarded as in the main belonging to it. The following are representative examples drawn from the recorded analyses before me. The animals were taken in the freshly-opened state:—

Oyster.		Sugar per 1000, expressed as glucose.	Cupric oxide reducing power of the sugar present in relation to that of glucose at 100.
Dutch native	{ before sulphuric acid { after ,, ,, ,,	1·230 2·330	} 52
Dutch native	{ before ,, ,, ,, { after ,, ,, ,,	0·730 1·580	} 45
Whitstable native	{ before ,, ,, ,, { after ,, ,, ,,	0·630 2·300	} 27
Cooking oyster	{ before ,, ,, ,, { after ,, ,, ,,	1·050 4·530	} 23
Mussel.			
Salt-water mussel	{ before ,, ,, ,, { after ,, ,, ,,	0·590 1·960	} 30

From what will be adduced further on (p. 194, *et seq.*), when reference is made to the sugar existing in other structures of the body, it will be seen that, taking the condition at the moment of death, no material difference is discernible between the liver and these other structures. In the spleen, for example, and the pancreas, kidney, brain, lung, placenta, and the same is true of the egg, sugar is invariably to be found; and, summarily expressed, the amount may be said to vary from about 1 or 2 to 3 or 4, and occasionally more, per 1000. In muscle the ordinary range stands rather higher, and quantities of 6 and 7 per 1000, and even beyond, are sometimes met with.

Looking, therefore, to the state existing at the moment of death, we find nothing, as far as sugar is concerned, to lead us to view the liver as standing in a different position from the other structures of the body. There is this, however, that is distinctive as regards the liver: the amount of glycogen, due to the position in which the organ is placed in relation to ingested carbohydrates, is, under normal conditions, much larger than elsewhere; and, in addition, there is present, or, it may be, there becomes developed, at death a very energetic sugar-forming ferment.

It is through the coexistence of these two factors that the capacity exists for the rapid and extensive production of sugar that is noticed at a suitably elevated temperature to occur. As far as the glycogen is concerned, there is the capacity during life, but obviously there must be a restraining or inhibitory influence in operation preventing the ferment change which occurs after death. It is not permissible to suppose that the same ferment change is taking place during life that is observed after death, and that the removal of the resulting sugar by the circulation constitutes the only difference existing; for, apart from other considerations elsewhere adduced in this work, which sufficiently negative such a proposition, the following argument stands in contravention.

From observation, it may be stated that sugar is produced in the liver during the first few—say ten—minutes after death to the extent of about 10 to 12 per 1000. With an average amount of glycogen present, even assuming the liver of the rabbit to be taken in which the amount is larger than in many other animals, it is shown by calculation (based upon a proportion of 50 per 1000) that the whole would disappear in about three-quarters of an hour if the production of sugar took place at the rate above mentioned. Further, in the case of the carnivorous animal, from the smallness of the amount of glycogen existing, it would sometimes happen that only a limited number of minutes would be required for a total disappearance to occur. It is true that whilst digestion and absorption are going on a formation of glycogen is taking place, which would have a counterbalancing effect. At a period of fasting, however, no such source of production exists, and yet, after a fast of twenty-four or even forty-eight hours, a considerable amount of glycogen is ordinarily found to be present, it being only after more prolonged fasting that it may be expected to be absent.

The train of reasoning that I have adduced does not stand upon a mere postulatory basis. Under certain conditions, which may be evoked experimentally, sugar is actually produced in the liver during life. As a result, it reaches the general circulation, and thence the urine, which thus becomes an indicator of the state of things existing within. Years ago I noticed, and was struck by, the short time sufficing for the liver to lose its glycogen in experiments attended with the artificial production of glycosuria. With animals killed at the end of an hour, or even less, I often failed to observe the presence either of sugar, by rough testing with Fehling's solution, or of glycogen. The inhalation of carbonic oxide, either directly, mixed with air, or indirectly, employed in the form of puff-ball smoke, constitutes one of the conditions leading to the production of sugar in the liver and its passage into the blood and urine. I have recently, with the improved methods of analysis now at command, conducted experiments of a quantitative nature. In these, the blood collected at the instant of death showed the presence of an abnormally large quantity of sugar, the amount standing in one instance as high as 4·38 per 1000. The livers were taken without subjection to special treatment, and thus attention required to be given to the amount of sugar present, as well as that of glycogen, in order that loss of the latter from *post-mortem* change might not escape consideration. In the case of a cat, submitted for a quarter of an hour to the influence of puff-ball smoke, the glycogen figures stood at 5·05 per 1000 and the sugar at 13·69 per 1000, both expressed as glucose. In that of another cat, submitted to similar treatment for half an hour, the glycogen figures stood at 3·10, and the sugar at 4·94 per 1000. In that of a dog, similarly treated for three-quarters of an hour, the figures for glycogen were 4·27 per 1000, and for sugar 1·05 per 1000.

The position pertaining to life appears to be an anomalous one, but in reality it may be said to be analogous to that which obtains in relation to the coagulation of the blood. Here the factors tending to produce coagulation are prevented normally during life from coming into play, but are permitted to do so after removal of the blood from the vessels. The analogy, even, may be followed further, for conditions may arise which permit the blood to coagulate whilst contained in the living vessels, and in like manner the liver may be placed under conditions to permit of the manifestation of ferment activity, resulting in the production of sugar, its passage into the circulation, and its escape with the urine.

Production of Sugar in the Liver after its Removal and the Passage of a Stream of Water through its Vessels to wash out the Blood.

It was discovered by Bernard, at an early period of his investigations, that in the liver-substance washed free from blood the capacity exists for the production of sugar. If means be adopted to connect the portal vein with a water-tap, and the tap be turned on, the water passes through the continuing vessels and escapes from the hepatic vein, carrying away the blood, and at the same time sugar and a certain amount of glycogen. During the process the liver becomes enormously swollen and œdematous, and likewise loses its colour. If it is subsequently placed in a position for ferment action to occur, sugar is found to be produced. Whilst such, from Bernard's time, has been known, nothing has hitherto been said about the nature of the sugar that is formed.

Upon the strength of the collection of results before me, the statement may be made that the liver-substance contains a ferment possessing a glucose-forming capacity, but that a variable kind of sugar product is met with.

Whilst a product with a cupric oxide reducing power more or less considerably below that of glucose is ordinarily encountered, glucose, on the other hand, may happen to be found. It has appeared to me as though there has sometimes been an extensive amount of glycogen transformed into a product of low cupric oxide reducing power, and at other times a less amount of material transformed, with the acquirement of a higher degree of cupric oxide reducing power.

The effect of adding blood to the washed liver is to increase materially the amount of sugar produced. As regards the nature of the sugar formed, nothing more definite can be said than that a considerable variation has been noticed in the cupric oxide reducing power of the product found to be present.

Production of Sugar in Liver Substance previously Coagulated by Alcohol.

It is liver-substance in a fresh state which has thus far formed the subject of consideration, and nothing is deducible from the information that has been supplied which can be taken as pointing to the production of sugar being other than the result of simple ferment action independent of connexion with vital activity: an action comparable to that exerted by the ordinary amylolytic ferments—diastase, ptyalin, &c. The only difference that is presented is that, in the one case, the capacity exists for carrying on the change to the stage of glucose, whilst, in the other, it only exists to a sufficient extent to lead to the production of maltose.

I now proceed to show that the liver-substance, after subjection to the coagulating influence of alcohol, still retains its capacity for sugar production—a fact which affords absolute proof, if indeed the idea could now exist in the mind of anyone that proof were wanting, that the phenomenon is not dependent upon the metabolic power resident in the living cells of the liver.

After coagulation by alcohol, the liver-substance is not prone to undergo change on keeping, and consequently may be preserved for any length of time in a condition convenient for the purposes of experiment in connexion with the question of ferment action. This question in its bearings upon the liver I have somewhat extensively studied, and in the succeeding pages I will give an account of the results obtained.

The liver selected for use should be one containing a fair amount of glycogen, and should be taken before loss of glycogen by transformation into sugar has been permitted to any marked extent to occur. After being thoroughly reduced to a pulp in a mortar, it is placed in a sufficiency of alcohol to fully secure coagulation. The alcohol is afterwards strained off, and with it is removed the greater portion of the sugar that may happen to have been present. The coagulated material is now allowed to dry, either spontaneously or over sulphuric acid in a desiccator. Heat must not be employed, unless the temperature be kept below the point at which the ferment becomes destroyed.

If the liver-substance thus prepared be treated with water and exposed to moderate warmth, change ensues, attended with the production of sugar. To reveal this sugar-production the employment of an analytical procedure is necessary, and a determination must be made, not only of the amount of sugar in the product at the end of the experiment, but also of the small amount existing in the dry material that escaped removal with the alcohol used in the preliminary step of coagulation: the figures for this requiring to be deducted from the others in order to obtain a representation of the actual amount of sugar produced.

The process I have mentioned as being in general use in my investigations supplies the information that is wanted. The sugar, after being fully extracted with alcohol, is estimated by titration with the copper test, before and after boiling with sulphuric acid. In this way both its amount and nature are revealed.

The residue from alcoholic extraction is, in the next place, subjected to boiling with potash and treatment with spirit to precipitate the glycogen present. The amount of this is subsequently ascertained by conversion into glucose by the agency of sulphuric acid and the employment of the copper test.

By means of this double process, data are supplied for comparing the gain of sugar with the loss of glycogen, and, if the carbohydrate is expressed throughout as glucose, we have equivalent terms of expression to deal with, permitting the figures on the two sides to be read off as they stand.

On experimenting in the manner that has been described, we learn that a progressive production of sugar takes place, concurrently with a disappearance of glycogen. In some of the experiments that I have conducted, the change has been observed to advance almost to the point of a complete replacement of the glycogen by sugar. We learn further, it may be said, that a gradual increase of the cupric oxide reducing power of the sugar produced takes place, until the stage of glucose is reached.

The following tabular representation shows the results obtained in an experiment where the product was examined after varying periods of exposure to a temperature of 38° C. 1·5-gram portions of the dried liver-substance with 20 c.c. of water were placed in separate flasks, and exposed alongside each other for the time specified in the table. They were then removed and submitted to analysis. The figures given to represent the state of the dried liver-substance existing at starting constitute the mean of four analyses, which, it may be remarked, stood throughout in close accord with each other.

Production of Sugar in Liver-substance, after coagulation by Alcohol. Results expressed as Glucose.

	Sugar, before treatment with sulphuric acid.	Sugar, after treatment with sulphuric acid (glucose).	Relation of the cupric oxide reducing power of the product to that of glucose.	Glycogen, expressed as glucose.
Liver-substance at once............	0·012 gram.	0·014 gram.	86 to 100	0·183 gram.
After 30 minutes at 38° C.	0·050 ,,	0·068 ,,	74 to 100	0·095 ,,
,, 1 hour ,, ...	0·070 ,,	0·093 ,,	75 to 100	0·080 ,,
,, 2 hours ,, ...	0·083 ,,	0·115 ,,	72 to 100	0·062 ,,
,, 4 ,, ,, ...	0·096 ,,	0·116 ,,	83 to 100	0·029 ,,
,, 6 ,, ,, ...	0·108 ,,	0·130 ,,	83 to 100	0·027 ,,
,, 7 ,, ,, ...	0·117 ,,	0·135 ,,	87 to 100	0·020 ,,
,, 7 ,, ,, and subsequently 18 hours at the ordinary temperature..	0·131 ,,	0·135 ,,	97 to 100	0·014 ,,

The figures in the columns under the headings of "Sugar, after treatment with sulphuric acid (glucose)" and "Glycogen, expressed as glucose," enable us to compare the gain of sugar with the loss of glycogen. Theoretically, on the assumption that glycogen is simply transformed into sugar, there should be an accord between the two sets of figures, or, in other words, no alteration in the total amount of carbohydrate. Practically, as seen in the table above, which agrees in the main with what is shown by other results that have been obtained, a general correspondence between gain and loss is to be observed, especially after limited periods of exposure. Upon the whole, however, it must be said that the loss is usually somewhat greater than the gain, and this becomes more marked as the experiment proceeds, ending with a pronounced diminution in the total carbohydrate where the exposure has been allowed to run on for a period of twenty-four hours.

It may here be remarked that experimental observations upon sugar-production in the fresh liver have yielded the same kind of evidence regarding the relation between gain of sugar and loss of glycogen. At the same time, it must be stated that in some instances results have been met with which can only, with our present knowledge, be characterised as presenting an anomalous appearance. It has seemed as though carbohydrate material has, upon some occasions, been brought into evidence from a latent or some other state, and, conversely, upon others, has disappeared from view.

With reference to the disappearance of carbohydrate, the experiments I have conducted in search of an explanation have led to the discovery of a point in connexion with the analytical procedure for the determination of glycogen, which goes towards accounting for a certain amount of loss. The product for analysis, it will be remembered, is first of all extracted with alcohol for the removal of the sugars. The coagulated residue containing the glycogen is then boiled with potash to disintegrate and dissolve the nitrogenous matter, and as far as possible place it in a position to be soluble in, and susceptible of removal by, alcohol. As long as glycogen is the principle that has to be dealt with, no sensible amount of destruction is occasioned by the boiling with potash, and, in harmony, it is noticeable that in the analyses of glycogen-containing products which have not been exposed to the modifying influence of ferment action the results are found to stand in close conformity with what might be looked for. Where products, however, are dealt with in which change through ferment action has occurred, a certain amount of loss of carbohydrate is, as an ordinary occurrence, observable in the analytical results obtained. Thus it is after, and only after, ferment action that the analytical results show the loss of carbohydrate.

Upon reflecting on the matter, the question presented itself to my mind whether, after the ferment action, there might not be a dextrin precipitated together with the glycogen by the alcohol, which failed to resist destruction during the process of boiling with the 10 per cent. solution of potash. I submitted the question to the test of experiment in the following way, and the results obtained furnished a decided answer.

For the settlement of the point, it is immaterial whether recourse be had to the employment of starch or glycogen, on account of the analogous positions held by these bodies in relation to the matter and the similarity of the products generated. It is also immaterial which of the amylolytic ferments is made use of. In the experiments performed, starch, paste, and pancreatic ferment were the materials employed.

After ferment action had been allowed to proceed for a short time, alcohol was freely added to dissolve out and separate the sugar and soluble dextrins formed. The residue was collected as in the ordinary analytical procedure, and, after being washed, was divided into two equal portions. Both were then mixed with the usual quantity of potash. Thus far, the steps adopted presented no variation from the usual course. In the next step, however, one portion was boiled for the accustomed time with the potash, whilst the other was simply placed in contact with it in the cold. Each was then poured into spirit, and the respective precipitates were afterwards collected and subjected to the inverting action of sulphuric acid. The results obtained showed a conspicuously smaller amount of carbohydrate where the product had been boiled with potash than where it had been simply treated with potash in the cold. Seeing that starch is not attacked by boiling with potash solution of the strength used (and the same holds good for glycogen), it is rendered evident that the residue remaining from the alcoholic extraction of the product of ferment action contained something besides untransformed starch. It may be assumed that some slightly transformed material, insoluble in spirit, existed, which, unlike the starch, failed to resist destruction on boiling with the 10 per cent. solution of potash. The loss of carbohydrate matter appearing in the results is thus to be accounted for without bringing into the question the occurrence of loss from the ferment action.

Alcohol-coagulated Liver-substance with Blood.

A tabular representation of the results of experiments bearing on sugar-production in alcohol-coagulated liver-substance taken alone was given a few pages back. I have conducted similar experiments upon the same specimen of liver-substance, with the addition of blood which had been dried at a temperature below that destructive of ferment activity, and the results, as in the case of those obtained from the admixture of blood with the fresh liver, show that a larger production of sugar takes place than when the liver is dealt with alone. A comparison of the table given below with that previously introduced stands in support of this statement. An effect also noticeable is that the product possesses a somewhat higher cupric oxide reducing power. The quantity of dried liver taken was the same as in the other experiments, viz. 1·5 grams, and the quantity of dried blood employed was 1 gram. The sugar intrinsically belonging to the blood, as shown by an examination made, was too insignificant in amount to need consideration.

Production of Sugar in Alcohol-coagulated Liver-Substance mixed with Blood. Results expressed as Glucose.

	Sugar, before treatment with sulphuric acid.	Sugar, after treatment with sulphuric acid (glucose).	Relation of the cupric oxide reducing power of the product to that of glucose.	Glycogen, expressed as glucose.
[Liver substance............	0·012 gram.	0·014 gram.		0·183 gram.]
Liver substance and blood				
After 30 minutes at 38° C.	0·062 ,,	0·074 ,,	86 to 100	0·091 ,,
,, 30 ,, ,, ...	0·068 ,,	0·088 ,,	84 to 100	0·077 ,,
,, 1 hour ,, ...	0·094 ,,	0·121 ,,	77 to 100	0·076 ,,
,, 2 hours ,, ...	0·124 ,,	0·152 ,,	78 to 100	0·042 ,,
,, 4 ,, ,, ...	0·121 ,,	0·143 ,,	82 to 100	0·015 ,,
,, 6 ,, ,, ...	0·121 ,,	0·133 ,,	85 to 100	0·013 ,,
,, 7 ,, ,, ...	0·144 ,,	0·158 ,,	91 to 100	0·015 ,,
			94 to 100	

Influence of Sodium Carbonate and Citric Acid on Ferment Change in the Liver.

I have tried the effect produced by small quantities of sodium carbonate and citric acid upon the change occurring in the alcohol-coagulated liver-substance which has just been dealt with as a basis of observation. The issue of the experiments conducted is to show that the influence of the presence of sodium carbonate is in the direction of diminishing the amount of transformation taking place, without in any marked manner affecting the cupric oxide reducing power of the sugar produced; whilst the influence of citric acid is in the direction of increasing the cupric oxide reducing power of the sugar-product, rather than in that of producing any decided alteration in the amount of carbohydrate transformed.

The effect of a large amount of sodium carbonate, like that of the caustic alkali, is to arrest ferment change. This may be shown not only by treatment of the liver after death, but likewise by the introduction of the agent into the organ during life. In an experiment upon a dog, placed and kept under the influence of ether, a lobe of the liver was isolated by a ligature and excised. 20 grams of sodium carbonate in 80 c.c. of water were then injected into a branch of the mesenteric vein. Death occurred immediately after the completion of the injection. A portion of the liver, which was shown by its black colour to have been fully penetrated by the agent, was taken for examination, as well as the lobe that had been excised before the injection was made. In each case, a certain period of time elapsed between the removal of the piece to be examined and the commencement of the analysis. The results obtained were as follows:—

		Sugar per 1000, expressed as glucose.
Portion of liver excised before the injection of the sodium carbonate	before sulphuric acid .... after ,, ,, ...	21·200 21·200
Portion of liver removed after the injection of the sodium carbonate	before ,, ,, ... after ,, ,, ...	1·600 2·200

In another experiment a dog was similarly anæsthetised. After the isolation and removal of one lobe of the liver, 10 grams of sodium carbonate in 40 c.c. of water were injected into the hepatic duct. Ten minutes later, the animal was killed by pithing. A portion of the liver that had been penetrated by the injection was, as in the preceding experiment, made the subject of comparative examination with the portion previously removed. As a further step, portions of the two specimens were placed aside and examined on the following day. The analyses yielded the following results:—

		Sugar per 1000, expressed as glucose.
Portion of liver excised before the injection of the sodium carbonate, taken a short time after removal..	before sulphuric acid after ,, ,,	9·400 10·000
Portion of the same specimen taken on the following day...	before ,, ,, after ,, ,,	13·800 14·000
Portion of liver removed after the injection of the sodium carbonate, and taken shortly afterwards.........	before ,, ,, after ,, ,,	1·200 2·000
Portion of the same specimen taken on the following day...	before ,, ,, after ,, ,,	2·000 2·800

It will be seen that the sodium carbonate employed in these experiments had the effect, practically, of arresting ferment action. The agent thus affords a means of placing the liver in a position to escape undergoing *post-mortem* change, and to permit of a representation of the state belonging to life being obtained without recourse to the precautionary measures otherwise necessary in the process of examination. If reference be made to the analyses in which *post-mortem* change was prevented by freezing (pp. 137—138), it will be found that a strict accord is noticeable in the figures yielded by the two modes of experimenting.

THE BLOOD IN RELATION TO SUGAR.

In accordance with what is observed elsewhere throughout the system, the blood is found to contain a certain amount of sugar. Both the nature and amount of this I will proceed to consider.

Nature of Sugar present in Blood.

To determine the nature of the sugar present the process of alcoholic extraction, previously described in detail (p. 61), should be adopted in order to avoid the simultaneous extraction of glycogen which, to a certain extent, occurs when water is employed, and which would lead to the introduction of error through the production of sugar under the process of treatment with sulphuric acid.

Observation, conducted upon blood derived from different sources, shows that the kind of sugar found in all parts of the circulation, with the exception of the portal system, possesses a cupric oxide reducing power that is not, as a rule, increased, and, if increased, only slightly so, by boiling with sulphuric acid—a character which implies, broadly speaking, the existence of glucose. This, I may state upon the strength of a very large number of observations, is the kind of sugar present under ordinary or natural conditions, but, as will be subsequently shown, under certain deviations from the ordinary state, as, for instance, after the administration of anæsthetics, the inhalation of carbonic oxide, &c., the sugar met with is usually one possessing a cupric oxide reducing power more or less below that of glucose.

158   THE BLOOD IN RELATION TO SUGAR.

Acicular glucosazone crystals from the blood of the general circulation
(rabbit). Magnified 400 diameters.

The blood of the portal system, however, differs from that of the other parts of the circulation, and, at the same time, resembles, as will be seen from the analyses to be given later on, the solid organs and tissues of the body. Its condition was fully referred to in a previous part of this work, and it was there shown that it ordinarily contains, not only after the ingestion of carbohydrate food, but likewise after animal food (and even in some instances the condition has been observed at a time of fasting), a sugar with a lower, and it may be a considerably lower, cupric oxide reducing power than that of glucose.

The blood, then, excluding that belonging to the portal system, contains sugar in the form, broadly speaking, of glucose, and in this respect it holds a position differing, so far as my observations have extended, from that existing in the other parts of the economy.

Amount of Sugar present in Blood.

Having spoken of the nature of the sugar in the blood, I will now pass to the consideration of its amount. Upon this point considerable diversity exists in the statements that have been made by different observers. The question of the amount of sugar naturally present in the blood is one that must be regarded as possessing more than an intrinsic importance on account of the meaning it may have in relation to considerations standing outside the simple question of fact appertaining to the blood itself. It may therefore be looked upon as a point upon which no doubt should be allowed to exist. The enquiry requires to be approached and carried out in a scrupulously guarded and careful way, but, with the adoption of proper measures, experience would lead me to say, the information wanted may be easily and reliably obtained. The manipulative process involves two distinct steps of procedure, each of which calls for the bestowal of close attention. Not only is it necessary that the method of analysis should be such as to yield reliable results, but also that the blood for analysis should be collected in such a manner as to afford a representation of the natural state.

Analytical proficiency in the early days of research in relation to the physiology of sugar in the animal system had not attained a position to permit of satisfactory quantitative determinations being made in an albuminous and coloured product like blood. Advance, however, has in this, as in so many other directions, taken place, and I think it may now be considered, with reference to the point in question, that a position has been reached leaving little or nothing to be desired.

Aqueous extraction of the sugar, with the adoption of appropriate measures for the separation of colouring and albuminous matters, if resorted to, will yield a liquid to which the ordinary process of sugar estimation may be applied, but, unless the nature of the sugar present is known, the result obtained will not supply the information required for the expression of the amount. In the absence of a knowledge of the nature of the sugar existing, it is necessary that the process of alcoholic extraction should be employed, in order that the sugar may be obtained free from glycogen, and so permit, after the usual treatment with sulphuric acid, of a truthful expression of its amount being given under the form of glucose.

When the process of extraction is carefully carried out, and the ammoniated cupric test is employed for the subsequent estimation of the sugar, the degree of accuracy attainable by experienced hands is such as almost to excite surprise, looking at the complex and coloured nature of the original product dealt with. In support of this statement, I may refer to the results given in an early part of this work, when the subject of analytical procedure in general was under consideration (p. 79). Examples are there introduced showing the close accord that is obtainable in the figures from duplicate analyses.

The conditions existing in connection with the collection of the blood for analysis constitute an item of consideration, it may be, of greater importance than the analytical procedure itself, inasmuch as neglecting to give due attention to the requisite precautions may give origin to greater error than would be likely to arise from a faulty analysis. How rapidly sugar may be adventitiously produced in the liver and lead to the presence of an abnormal amount in the blood has been already shown, and in obtaining a specimen for examination fallacy from such a source must be avoided by the collection being made under conditions to secure a representation of the natural state.

This matter, as is known, I drew attention to upwards of thirty years ago. In a communication published in the Transactions of the Royal Society for 1860, I showed that the condition which had been previously taken as representative of the natural state, widely differs in reality from it. With the occurrence of death, an alteration in the amount of sugar present in the blood speedily ensues, but this had not hitherto been recognised and taken into account. Thus, through the *post-mortem* being regarded as expressive of the *ante-mortem* state, a false foundation existed for the physiological doctrine which happened to be constructed upon what had been observed. At the outset a qualitative examination was all that the resources at command permitted to be made, and what was considered to be the characteristic reaction to be looked for from the blood of the inferior cava and right side of the heart was one attended with a plentiful reduction of the copper test. The amount of sugar now revealed by quantitative determination as actually existing in the blood collected from these parts in an ordinary way after death, depends upon the extent to which it may happen to have been influenced by the *post-mortem* production of sugar in the liver. A varying quantity, from about that naturally belonging to the living state to one largely exceeding it, may be found.

The point under consideration is well illustrated by the results obtained in a recently conducted experiment, of which the following are the particulars. A dog was suddenly killed by pithing, and the chest immediately opened. The heart was then grasped at its base with the hand, and a ligature applied so as to prevent any further ingress or egress, and thus permit of the blood contained in its cavities close to the moment of death being obtained for examination. By an incision above the ligature, the heart was excised, and the blood which it contained subsequently collected. With the excision of the heart, the blood present in the large vessels of the system flowed into the chest. This was dipped out, and a portion taken for analysis. The following are the results that were yielded by the respective specimens.

Sugar per 1000, expressed as glucose.

Blood from the cavities of the heart	before sulphuric acid			0·850
	after	,,	,,	0·876
Blood collected from the chest	before	,,	,,	2·180
	after	,,	,,	2·237

The amount of sugar found in the blood of the general circulation under a natural and tranquil state of the system presents but little variation, and, as will be seen from the details to be introduced later on, is less than what is met with in the solid organs and tissues. From a collection of upwards of 100 observations, conducted at various times over a number of years, upon the dog, cat, rabbit, sheep, ox, horse and pig, THE AMOUNT OF SUGAR NATURALLY PRESENT IN THE BLOOD OF THE GENERAL CIRCULATION MAY BE STATED TO RANGE from ABOUT 0·6 TO 1·0, OR A LITTLE OVER 1·0, PER 1000.

Most of the determinations upon which the above statement is based were made with the ammoniated cupric test. In many, however, the gravimetric process described at p. 69 was employed. By this process the copper thrown down from the Fehling's solution used is collected, dissolved, and subsequently deposited by galvanic action upon a platinum cylinder for weighing. Although there are sufficient grounds for considering that full reliance may be placed upon the information supplied by the ammoniated cupric test, it is nevertheless satisfactory to have it corroborated by a different kind of analytical procedure; and further so when, as illustrated by the examples given at p. 80, the accuracy of both processes is confirmed by the close accord in the figures obtainable under their application to the same specimen of blood.

The blood, then, of the general circulation possesses a constitution in relation to sugar presenting but little variation whilst ordinary or natural conditions exist. Disturbed states of the system, however, quickly occasion the presence of an increased amount, but the kidneys perform an eliminative office like that which they discharge in the case of urea. Under any increase of sugar in the circulation, the urine is immediately influenced to a proportionate extent, and the escape of sugar thus occurring constitutes a provision for keeping down accumulation, as is exemplified by the elimination that is observed to take place in diabetes.

In former times, when less facility existed for the quantitative determination of the sugar of the blood, the amount corresponding with what I have represented as naturally present during life was usually spoken of as a "trace," the expression being founded upon the slight reaction yielded by the copper test. I do not think previous attempts had been made to determine what this "trace" meant, expressed in figures, but in my communication contained in the Transactions of the Royal Society for 1860, I gave the results I had obtained from the analysis of three specimens of the blood of dogs, and stated that the quantities stood respectively at 0·470, 0·730, and 0·580 per 1000. These amounts, it will be seen, although obtained upwards of thirty years ago, coincide closely with those yielded by the present improved methods of analysis.

I consider there ought not to be any disagreement, certainly at the present time, about the amount of sugar present in blood; and, should a discordance exist, it is more likely to depend upon something connected with the collection of the blood than upon the analytical procedure. The figures given by some observers stand in close accord with, and by other observers even a little lower than, my own. A striking want of uniformity is noticeable in the results obtained by Bernard, and, whilst placing the lowest point of range at 1 per 1000, he says ('Comptus Rendus,' 1876, p. 1409) that in the normal state the sugar varies from 1 to 3 per 1000. Seegen represents the amount as oscillating between 1 and 2 per 1000, and remarks that it is only exceptionally that it exceeds 2 per 1000. The state of the urine suffices, I consider, to show that these ranges of

Bernard and Seegen cannot be otherwise than founded upon error. The larger quantities are incompatible with the state presented by healthy urine. They mean, as is rendered evident by reference to the table at p. 192, representing the condition of the blood in cases of diabetes, the existence of a state that would be attended with the presence of a large amount of sugar in the urine.

A contrast exists between the blood of the portal system and that of the general circulation. Whilst the amount of sugar in the former is in a direct manner influenced by the ingestion of carbohydrate matter, no such influence is exerted upon the contents of the general circulation.

I have mentioned that precautions must be observed in collecting the blood for examination, in order to obtain a representation of the natural state. If the collection be made from an artery or vein during life, the animal must be at the time in a state of tranquility. Obstruction of the breathing, violent struggling, and the administration of an anæsthetic suffice, through their disturbing effects upon the system, to produce an increase of sugar in the blood.

Bernard noticed that the blood of the jugular vein of an animal contained more sugar after muscular efforts provoked by holding the nose for a short time so as to impede the respiration than it did before. I also, in an experiment performed some years ago, found that blood taken from a dog which had been subjected to obstruction of the breathing for half-an-hour possessed a strongly saccharine character, and, in another instance, where the experiment was carried on for a longer period, that the urine even had acquired a saccharine impregnation sufficient to give a strong reaction with the test for sugar. In like manner, after the administration of an anæsthetic, the urine may be found to give even a fairly strong saccharine reaction with the copper test. I took the opportunity at one time to institute observations upon the urine of patients in Guy's Hospital, before and after the inhalation of chloroform, and found in every instance a decided effect, in some even a strong effect, produced in the direction mentioned.*

In obtaining blood from the killed animal for examination, the

---

* The cupric oxide reducing effect of urine after the administration of chloroform is now stated to be due to glycuronic acid, a body closely allied to and supposed to be derived from dextrose, but presenting a rather higher state of oxidation.

collection must be made before time has been allowed for its constitution in relation to sugar to become altered from that appertaining to life.

In the laboratory the conditions are under the command of the experimentalist, and the method of killing employed should be one to occasion death as instantaneously as possible. After the destruction of life, say by pithing, a scalpel, or pair of scissors, should be instantly thrust into the chest, and the heart and large vessels freely incised. The chest being then quickly opened, the required quantity of blood should be at once dipped out or otherwise procured for examination.

At the slaughter-house different methods of killing are adopted. The method to be chosen for securing the attainment of the object in view, is one in which death is occasioned purely by hæmorrhage. In the case of the sheep, the neck is pierced with a knife, and the blood drained off by severance of the vessels. The first portion of the blood that escapes is the proper specimen for examination. The bullock slaughtered by the Jewish method affords another instance of killing by hæmorrhage. By a single incision with a long, sharp knife, all the soft structures of the neck in front of the vertebral column are divided. An immense gush of blood from the arteries takes place, and, as in the case of the sheep, the first portion should be collected.

In the course of an inquiry I was some time ago conducting, I was led to submit to examination specimens of blood from the cut bullock obtained, one at the first instant of its escape, and the other a few moments later. The specimens were subjected to the gravimetric process of analysis, and, as will be seen from the subjoined figures, a slight difference was found to exist between the two in the direction of a larger amount of sugar in the second portion, showing how the lapse of an apparently insignificant amount of time suffices to bring about an altered condition.

Blood of Bullock killed by Jewish Method, collected at the Commencement and at a later Period of the Flow. Gravimetric Process of Analysis.

Sugar per 1000 parts.

	First collection.		Second collection.	
	Duplicate analyses.	Mean.	Duplicate analyses.	Mean.
Observation 1......	0·544 0·555	0·549	0·581 0·585	0·583
Observation 2......	0·629 0·620	0·624	0·673 0·686	0·679

After the employment of the pole-axe method of slaughtering higher figures are ordinarily obtained, and, concordantly, the method involves the lapse of a longer time before the blood is withdrawn. The animal is first felled by the blow of the axe, and through the opening made a cane is then passed down the spinal canal to crush the medulla oblongata and spinal marrow, a proceeding which not only effectually kills outright, but puts a stop to the occurrence of reflex movements. An incision is next made into the lower part of the neck, and the knife passed on into the chest so as to penetrate the superior vena cava, or possibly the right auricle of the heart, and permit the blood to escape. Subjoined are the results derived from the analysis of four specimens obtained in this way. In connexion with these specimens the remark is recorded that instructions were given to my assistant to get the incision and collection made with as little loss of time as possible after the felling of the animal. In two of the observations the figures happen to about agree with those obtained under the methods of slaughtering previously referred to. In the other two, however, the higher kind of figures are presented.

Blood from the Bullock Slaughtered by the Pole-axe. Gravimetric Process of Analysis.

Sugar per 1000 parts.

	Duplicate analyses.	Mean.
Observation 1........	0·595 0·597	0·596
Observation 2........	0·655 0·662	0·668
Observation 3........	1·037 1·070	1·053
Observation 4........	1·091 1·097	1·094

Arterial and Venous Blood in relation to Sugar.

Bernard made the statement* that a larger amount of sugar is present in arterial than in venous blood, and the inference has been drawn that a disappearance, occurring as a result of the functional disposal of carbohydrate matter in the system, takes place during the transit from the arterial to the venous system. Considerable discordance, however, is to be observed in the experimental results upon which the statement has been based, and I notice an instance amongst them in which an identity existed in the amount of sugar found in the blood of vein and artery. According to the particulars given of the experiment, a first collection of blood was made, from which the figures obtained were 1·250 per 1000 for the venous, and 1·480 per 1000 for the arterial, blood. A second collection was afterwards made, when it was found that the figures stood at 1·560 per 1000 for both the venous and arterial blood. In the instance where the largest amount of difference was observed, it appears from the details furnished that the experimental procedure adopted was as follows: Blood was withdrawn from the right crural vein, and immediately afterwards the corresponding artery was opened and blood collected from it. The results derived from the analysis of the two specimens are stated to have been 0·730 per 1000 for the venous, and 1·450 per 1000 for the arterial, blood, representing a difference of 0·720 per 1000 between the two—a difference, that is to say, about equivalent to the whole quantity of sugar which, according to my own observations, is found naturally to exist in the blood of the dog, the animal upon which the experiments were conducted.

The question that is being considered is one of great importance on account of the reasoning that may be based upon it. It happens at the same time to be one in connexion with which fallacy may most easily creep in through the difficulties attending the experimental procedure. Owing to the readiness and rapidity with which the blood becomes charged with an increased amount of sugar, as an effect of disturbed states of the system, error is certain to arise, unless the most scrupulous and guarded care is exercised with regard to the manner in which the collection of the respective specimens of blood

---

* Bernard, 'Comptes Rendus,' Tome lxxxiii, No. 6, p. 373; and 'Leçons sur le Diabète.' Paris, 1877.

is made. To obtain a truthful representation of the natural relative state of arterial and venous blood, the specimens to be compared must be collected absolutely at the same instant. Should the collection happen not to be absolutely simultaneous, the position is open for the occurrence of an alteration in the amount of sugar contained in the entire circulation, and this naturally would be read off as representing a difference in the relative amount present in the blood of the artery and vein. It is further desirable that before the collection of blood is made the animal should have been allowed to recover from the influence of the anæsthetic, and the effect of any struggling that may have attended the operation, so that a settled state of the contents of the circulatory system may have become established.

With a full realisation of the rigorous care and attention requiring to be bestowed upon all the details of the enquiry, I conducted the experiments which are to be found recorded in part in the 'Proceedings of the Royal Society' for 1877 (vol. 26, p. 346), and in full in my Croonian Lectures "On Certain Points connected with Diabetes" (pp. 71 *et seq.*), delivered at the Royal College of Physicians, in 1878, and published in the same year. In each of the experiments chloroform or ether was administered for the operative procedure of exposing the vessels, and the vessels chosen were the carotid artery of the one side, and the jugular vein of the other. After their exposure, and the required isolation from the adjacent tissues, a ligature was passed underneath them, and a knot tied in such a manner as to leave a loose loop surrounding them of sufficient length to serve for subsequently drawing them forward without giving pain to the animal, and thus without occasioning any struggling or disturbance. A period of an hour and a-half or two hours was allowed to elapse after this operative procedure for the system to recover from the influence of the anæsthetic. At the end of this time, without the employment of any forcible restraint, and whilst the animal was seated quietly on a table, the vessels were drawn forward, and blood collected simultaneously from each. The process of analysis was then at once commenced, and in the seven observations conducted the following results were obtained.

Comparative Examination of Arterial and Venous Blood collected simultaneously from the Dog during Life. Gravimetric Process of Analysis.

Sugar per 1000 parts.

	Arterial blood.		Venous blood.	
	Duplicate analyses.	Mean.	Duplicate analyses.	Mean.
Observation 1......	{0·806 / 0·817}	}0·811	{0·808 / 0·788}	}0·798
Observation 2......	{0·873 / 0·854}	}0·863	{0·896 / 0·863}	}0·879
Observation 3......	{0·918 / 0·948}	}0·933	{0·918 / 0·914}	}0·916
Observation 4......	{0·870 / 0·899}	}0·884	{0·859 / 0·873}	}0·866
Observation 5......	{1·079 / 1·081}	}1·080	{1·102 / 1·096}	}1·099
Observation 6......	{1·231 / 1·232}	}1·231	{1·240 / lost}	}1·240
Observation 7......	{1·155 / 1·170}	}1·162	{1·180 / 1·187}	}1·183

On casting the eye through the above list, it will be noticed that a close conformity exists in the figures for the corresponding specimens of arterial and venous blood. In some, the difference presents an excess on the side of the arterial blood; in others, on that of the venous. In each case, it does not amount to more than may be legitimately considered as falling within the limits of variation arising from the analysis, for it is not claimed that *absolute* accuracy is attainable. It must be borne in mind, with reference to the matter, that about 20 grams of blood were taken for analysis, and that the figures represent parts per 1000, so that what may be an exceedingly slight error in the analysis becomes magnified about fifty times in the multiplication employed for the expression of the result. A difference, for instance, of one-tenth of a milligram in the actual analysis would become a difference of 0·005 in the result expressed. The duplicate analyses give a trustworthy character to the evidence, beyond what it would otherwise possess, and the closeness noticeable in the counterpart results affords strong testimony of the precision attainable by the analytical process employed. In observation 6 one of the results for the venous blood is missing, owing to an accidental loss of some of the precipitated suboxide having occurred in the performance of the analysis. In observations 5, 6, and 7 the amount of sugar encountered stands higher than what is found to exist under ordinary circumstances. This, it may be considered, is attributable to the effect of the anæsthetic. If blood be withdrawn whilst the animal is actually under the influence of the anæsthetic it is possible that upwards of 2 parts per 1000 may be found, and, from the results referred to, it would appear that even after the lapse of an hour and a half or two hours the sugar may not have fallen to its standard amount.

Looking now at the results, taken altogether, we find that the mean amount of sugar, given by calculation from the mean figures furnished by the seven observations, stands at 0·995 per 1000 for the arterial blood, and 0·997 per 1000 for the venous. The difference between the two thus amounts to 0·002 per 1000, and it happens, as is seen, that the excess, such as it is, falls on the side of the venous blood.

By another mode of experimenting that suggested itself to me as open for employment, blood was obtained from artery and vein as expeditiously as possible after death, so as to anticipate the altered state induced by the *post-mortem* production of sugar. Four such experiments were performed upon dogs. The animal was in each case pithed, and, instantly afterwards, a scalpel was drawn across the artery and vein determined upon, without any attempt being made to isolate them. The blood that simultaneously flowed from the respective vessels was collected in capsules. The danger to be guarded against in the experiment was from the possibility that the flow might not be precisely even after the first gush, and that, in carrying on the collection, a little blood might be included which had undergone a slight degree of *post-mortem* modification. Attention was therefore given that the collection was not carried further than the withdrawal of the quantity absolutely required for the purpose of examination.

In observation 1, the collection was made from the jugular vein of one side of the neck, and the carotid artery of the other. In consequence of the experience gained in this experiment the crural artery was substituted for the carotid in the other three. Its more superficial position rendered it more accessible for division. The jugular was still retained as the vessel for yielding the venous blood. Since the blood throughout the arterial system possesses at a given moment the same constitution, it is obvious that any artery that is most conveniently situated for the performance of the experiment may be taken. In connection with observation 4, it is recorded that a little difficulty occurred in collecting the arterial blood, through its flowing slowly from the divided vessel, and that the last portion was dark in colour. In this condition it may possibly have just commenced to be influenced by the *post-mortem* influx of sugar from the liver.

Comparative Examination of Arterial and Venous Blood, collected simultaneously, instantly after Death. Gravimetric Process of Analysis.

Sugar per 1000 parts.

	Arterial blood.		Venous blood.	
	Duplicate analyses.	Mean.	Duplicate analyses.	Mean.
Observation 1......	0·938 / 0·915	0·926	0·904 / 0·897	0·900
Observation 2......	0·799 / 0·791	0·795	0·793 / 0·791	0·792
Observation 3......	0·849 / 0·847	0·848	0·847 / 0·854	0·850
Observation 4......	0·812 / 0·830	0·821	0·797 / 0·798	0·797

The results derived from these four observations present, it is noticeable, a less even character than those belonging to the previous set. The mean amount of sugar given for the arterial blood is 0·847 per 1000; and for the venous blood 0·834 per 1000. The difference stands at 0·013 per 1000, the higher figures being furnished by the arterial blood.

Placing now the eleven observations together we get 0·941 per 1000 as the expression of the mean amount of sugar found in the arterial blood, and 0·938 per 1000 in the venous. It thus appears, for the whole of the observations, that the sugar in the arterial blood exceeded to the extent of 0·003 per 1000 that in the venous.

The conclusion, I consider, may be drawn from the foregoing experiments that no material difference exists in the amount of sugar present in arterial and venous blood. As a corollary, it follows that no support is given to the view that carbohydrate matter in the form of sugar is allowed to reach the general circulation, for conveyance as a functional operation to the systemic capillaries to be disposed of by the tissues.

Professor Seegen, who widely differs from me upon other points, is, I find, in accord with me upon the one which has just been considered. In the French translation, which is before me, of his recent work on animal glycogenesis,* he says "Il resulte de là que, contrairement à l'opinion de Cl. Bernard et de Chauveau et d'accord avec les recherches de Pavy et autres, il n'y a pas de différence appreciable entre le sang artériel et le sang veineux quant à la proportion de sucre qu'ils renferment."

Blood after Withdrawal, in Relation to Sugar.

I have now finished what I have to say upon the question of the disappearance of sugar from circulating blood, and I will proceed to consider the question of its disappearance from blood after withdrawal from the system.

In correspondence with the statement that was made with regard to circulating blood, it was further asserted by Bernard† that a disappearance of sugar takes place in drawn blood, sufficient in extent and rapidity to furnish additional support to the theory of carbohydrate disposal to which attention has above been given.

At quite an early period of my investigations, and long before the matter which is being discussed was broached, I gave attention to the subject of the disappearance of sugar from drawn blood, and, from what I observed, I mentioned that a disappearance occurred, which seemingly ran concurrently with the changes that lead on to decomposition. I further mentioned that I had observed, with blood containing much sugar, an acid reaction produced, which I suggested was probably attributable to the formation of lactic acid. I also noticed that in the presence of fibrin and corpuscles the disappearance took place with greater rapidity than when serum alone was dealt with.

When the assertions of Bernard to which I have been directing

* 'La Glycogenie animale,' by Professor J. Seegen. Translated by Dr. Hahn : Paris, 1890, p. 100.

† Bernard, 'Comptes Rendus,' 19th Juin, 1876, p. 1406.

attention were made, I not only conducted the experiments I have referred to in relation to circulating blood, but also carried my investigations to drawn blood; and, in the prosecution of the enquiry, I was now aided by the methods of quantitative analysis which were not at my command when my original observations were undertaken. The experiments are to be found recorded in the 'Proceedings of the Royal Society' for June, 1877 (Vol. XXVI, p. 346), and April, 1879 (Vol. XXVIII, p. 520).

In giving a representation of what was found, I will first cite the particulars of five observations in which the blood was allowed to stand for moderate lengths of time at the ordinary temperature. The observations, it will be noticed, were conducted at different periods of the year, and thus under the existence of somewhat different temperatures.

Blood after Standing, compared with Blood taken Immediately. Gravimetric Process of Analysis.

	Sugar per 1000 parts. Mean of two analyses.	Percentage loss of sugar.
January 29th.		
Taken immediately............	0·786	—
,, after 1 hour............	0·739	6·0
April 25th.		
Taken immediately............	0·700	—
,, after 1 hour............	0·670	4·3
May 18th.		
Taken immediately............	0·766	—
,, after 1 hour............	0·751	2·0
,, after 23 hours	0·285	62·8
May 24th.		
Taken immediately............	0·786	—
,, after 1 hour............	0·728	7·4
,, after 24 hours............	0·302	61·6
May 26th.		
Taken immediately............	0·921	—
,, after 1¾ hours.........	0·793	13·9

In another set of experiments blood was allowed to stand for longer periods of time. In these the sugar was determined by the ammoniated cupric test, and not by the gravimetric process, which is only adapted for the examination of blood in a fresh state. With the occurrence of decomposition, the ammonia generated interferes with the deposition of the cuprous oxide, and thus in the case of the gravimetric process gives rise to a vitiated result. In the case, on the other hand, of the ammoniated cupric test, which in its principle of operation depends upon the solvent action upon the cuprous oxide exerted by ammonia, decomposition produces no vitiating effect and accordingly stands as no barrier to its employment. By this method of analysis, therefore, the examination can be conducted at any period, no matter whether the blood be in a fresh or a decomposed state. In these experiments the product for titration was obtained by the sulphate of soda method (*vide* page 59) resorted to for the gravimetric process, that is by boiling with sulphate of soda to coagulate albuminous and colouring matters, filtering, thoroughly washing the coagulum, and bringing the liquid to a known volume. The following examples may be selected and given to represent the character of the results obtained.

Blood after prolonged Standing at the ordinary Temperature compared with Blood taken Immediately. Sugar determined by the Ammoniated Cupric Test.

	Sugar per 1000 parts.	Percentage loss of sugar.
Blood of bullock—		
Day of withdrawal	0·775	—
1 day afterwards	0·334	56·9
2 days ,,	0·253	67·4
5 ,, ,,	0·231	70·2
Blood of bullock—		
Day of withdrawal	1·111	—
1 day afterwards	0·717	35·5
2 days ,,	0·545	51·0
3 ,, ,,	0·294	73·5

Blood with added Glucose subjected to prolonged Standing at the ordinary Temperature.

	Sugar per 1000 parts.	Percentage loss of sugar.
Blood of bullock—		
In fresh state	0·776	—
After addition of sugar	3·636	—
On the following day	2·811	22·7
On the third day	0·296	91·9
On the sixth day	0·228	92·7
The original blood, to which no sugar had been added, examined on the sixth day	0·225	70·0

In other experiments the blood was exposed to a temperature about equal to that of the body, and at the same time was subjected to the influence of currents of different gases. In order that the results might be expressed by larger figures, and any effect produced thus be rendered more visible, some glucose was added at the commencement of the experiment.

Blood with added Sugar exposed to a moderately elevated Temperature and to the Influence of Currents of different Gases.

	Sugar per 1000 parts.	Percentage loss of sugar.
Sheep's blood in a fresh state, with added sugar—		
Taken at once	1·850	—
After standing 7 hours at a slightly raised temperature	1·550	16·2
After the passage of oxygen for 7 hours at a slightly raised temperature	1·525	17·6
After the passage of carbon dioxide for 7 hours at a slightly raised temperature	1·525	17·6
After the passage of hydrogen for 7 hours at a slightly raised temperature	1·570	15·1

	Sugar per 1000 parts.	Percentage loss of sugar.
Sheep's blood in a fresh state, with added sugar—		
Taken at once................	1·634	—
After standing 2¼ hours at 38° C.	1·459	10·7
After the passage of oxygen for 2¼ hours at 38° C.	1·285	21·4
After the passage of carbon dioxide for 2¼ hours at 38° C. ...	1·100	32·7
Sheep's blood in a fresh state, with added sugar—		
Taken at once................	1·667	—
After standing 7 hours at 38° C.	1·342	19.5
After the passage of oxygen for 7 hours at 38° C.	0·992	40·5
After the passage of carbon dioxide for 7 hours at 38° C. ...	1·042	37·5
Sheep's blood in a fresh state, with added sugar—		
Taken at once................	1·475	—
After standing for 6¼ hours at 38° C......................	1·324	10·2
After the passage of oxygen for 6¼ hours at 38° C.	1·134	23·2
After the passage of carbon dioxide for 6¼ hours at 38° C...	1·209	18·0
Sheep's blood in a fresh state, with added sugar—		
Taken at once................	1·775	—
After standing 6¼ hours at 38° C.	1·567	11·7
After the passage of oxygen for 6¼ hours at 38° C.	1·475	16·9
After the passage of carbon dioxide for 6¼ hours at 38° C...	1·492	15·9
Sheep's blood in a decomposed state, with added sugar—		
Taken at once................	1·324	—
After standing 6 hours at 38° C.	0·667	49·6
After the passage of oxygen for 6 hours at 38° C.	0·606	54·2
After the passage of carbon dioxide for 6 hours at 38° C....	0·654	50·6

Looked at in their entirety, these experimental results show that an advancing disappearance of sugar takes place in drawn blood. Beyond this nothing definite can be said, and nothing I consider is deducible to warrant the conclusion that what is noticed to occur can be interpreted as having any physiological bearing. With an unstable body like sugar existing in contact with a complex organic product like blood, it is nothing more than might be expected that such disappearance as is observed should occur, and no need exists to bring into the question the operation of living action.

In the experiments where blood was subjected to the influence of the transmission of currents of different gases, it is noticeable that a greater disappearance of sugar occurred in the specimens that were so treated than in the counterpart specimens which were simply allowed to remain at rest. It is not, however, discoverable that more effect was definitely produced by one gas than by another, and it may be considered probable that the greater loss attending the employment of the gas was due to the physical effect of the molecular movement occurring.

The largest loss is observable where the blood employed was in a decomposed state. This is only in accord with what might be looked for.

A point to be noted in connection with the observations is that the blood invariably retained a certain amount of cupric oxide reducing power. In no instance, even where evidence of advanced decomposition existed, did the reducing power of the product fall below from 0·2 to 0·3 per 1000, expressed as glucose. What, in short, occurred was that the reducing power fell until the point named was reached, after which no further change ensued, however long the blood was kept. From this, the conclusion suggests itself that the blood contains a certain amount of something besides the sugar possessing cupric oxide reducing power, which, unlike sugar, resists the destructive influence of the decomposing matter around. If this is the case it follows that the sugar actually existing in the blood is less by 0·2 to 0·3 per 1000 than what has been represented under the assumption that the cupric oxide reducing effect observed is due solely to sugar.

Lépine's Theory regarding Glycolysis in Blood.—From the results of observations on the disappearance of sugar in drawn blood, Lépine has founded an argument in which he contends for the occurrence of glycolytic ferment action in the blood during life. Much discussion has recently taken place upon this matter. The view is of a nature to be likely to attract attention, seeing that it aims at accounting for the destruction of the sugar assumed under the glycogenic doctrine to reach the general circulation—a necessary provision for obviating the existence of diabetes as a general state. Weighty objections have been raised by various authorities to the experiments and conclusions of Lépine, looked at upon their own merits. There is, however, the further point, which has not been considered by others, that if, as follows from what has been adduced in this work, sugar does not reach the general circulation as the glycogenic doctrine implies, the *raison d'être* of a glycolytic ferment in the blood does not exist, and the question about glycolysis in the blood is devoid of the physiological significance that would otherwise belong to it.

THE URINE IN RELATION TO SUGAR.

Presence of Sugar in Healthy Urine.

I have incidentally referred to the elimination of sugar that takes place from the blood through the medium of the urine. I will now proceed to fully consider the relationship existing between blood and urine with regard to sugar. Whilst it is universally admitted that the blood contains a certain amount of sugar, the urine is in ordinary language spoken of as free from it. It is true that on testing a specimen of normal urine with Fehling's solution—the sugar test in common use—no reaction is obtained. This must not, however, be taken as proving the complete absence of sugar. There is a limit to the sensitiveness of every test, and in the case of sugar the amount present may be too small for the test to reveal it. Moreover, with urine, other materials are present which exert an influence upon the test contributing in some measure to mask the indication of sugar. Observation shows that an amount of sugar distinctly recognisable in a pure aqueous solution may not be recognisable when existing in urine. In water there is nothing to interfere with the deposition, and consequently the ready perception, of any suboxide that may be formed. In urine, on the other hand, the ammonia developed by the action of the potash of the test upon the nitrogenous matter present may suffice to hold in solution the suboxide produced when only small in amount. In may thus happen that no precipitate at all is produced, when in reality a certain amount of reduction has taken place. The only actual effect under the circumstances to be perceived is a certain extent of fading of colour, which it may be is so slight as to be barely appreciable, from the diminution of the oxide of copper by its conversion into suboxide. It is of course only where the quantity of sugar present is minute that it is liable to be concealed in this way. Where the amount is large the suboxide produced is beyond the capacity of the ammonia generated to dissolve it, and it then falls as the well-known precipitate belonging to the action of the test.

Although it may thus be found that no reaction is perceptible on testing healthy urine with the copper solution, the presence of sugar is nevertheless susceptible of being demonstrated by appropriate means. Brücke, it is known, some years back devised a process for the separation of sugar by throwing it down in combination with oxide of lead. The process is an eminently satisfactory one, and by its means sugar, to however minute an extent it may be present, can be extracted and placed in a position to be easy of recognition. By operating upon a sufficiently large quantity of urine, an amount of sugar is obtainable which admits not only of recognition but likewise of quantitative determination. I will here give the main points of the process, and for full details upon the whole subject may refer to an article written by me "On the Recognition of Sugar in Healthy Urine," which was published in the 'Guy's Hospital Reports' for 1876.

The urine is first treated with neutral followed by basic acetate of lead. The effect of this is to lead to the precipitation of the uric acid, sulphuric acid, phosphoric acid, hydrochloric acid, and, doubtless, some other constituents of the urine, as lead compounds, the sugar remaining in solution. Filtration is performed, and the filtrate is treated with ammonia, and a further quantity of acetate of lead, unless a large excess of it has been in the first place added. The sugar falls in the precipitate now produced, through the formation of a definite, insoluble compound of sugar and oxide of lead. The precipitate is then collected, and, in order to secure the complete removal of ammonia, washed until the washings no longer produce any effect upon reddened litmus paper. The next step is the liberation of the sugar from the compound existing in the washed precipitate. This may be effected by the agency of hydrochloric, sulphuric, or oxalic acid, but a somewhat coloured product is the result, and it has been suggested that the possibility may exist of the formation of sugar by the action of the acid upon some other constituent of the urine carried down with the lead precipitate. A more colourless, and therefore better, product is yielded by sulphuretted hydrogen. Although its application involves the occupation of considerable time, it constitutes decidedly the preferable agent for employment. The washed lead precipitate is subjected to the action of a stream of the gas until the sugar compound is completely broken up, which may be assumed to have occurred when the product has acquired an absolutely uniform black colour throughout. Filtration is next performed, and the excess of sulphuretted hydrogen expelled by heat. The liquid is then reduced to a small bulk, either over the water-bath or in the vacuum of an air-pump.

The product thus obtained is in a strongly acid state, and presents a considerable amount of colour. Neutralisation with carbonate of soda leads to the gradual deposition of a coloured substance, which should be removed by filtration. If any evaporation by heat is subsequently resorted to, the liquid should first be made slightly acid with acetic acid to guard against any destruction of sugar through the influence of alkalinity. To place the product in a more suitable position for the application of chemical tests, colour should be still further removed by the agency of animal charcoal, the charcoal employed having been well purified from lime. It must, however, be remembered that animal charcoal not only absorbs colouring matter but tends also to take up and hold sugar with some tenacity. Thorough washing is therefore required in order to fully recover the sugar.

As stated in the article ("Recognition of Sugar in Healthy Urine") to which I have referred, I applied the process that has just been described in outline, and satisfied myself of the truth of the assertion that in the product yielded sugar is to be found. I spared no pains to obtain adequate quantities of the isolated material to operate upon. The process of extraction was carried on from day to day for a considerable time, as fresh portions of urine were obtained, and altogether, in the course of the numerous observations undertaken, fully 100 litres passed under examination, in quantities of two or three litres at a time. Attention was given to secure that all the urine was derived from healthy persons, and the precaution was further taken of examining each collected portion with Fehling's solution, and rejecting any that gave the slightest indication of a reaction.

Subjected to examination by the application of the undermentioned tests, the product behaved in the following manner:—

Boiling with a solution of potash was attended with the production of the deep brown colour known to occur in the presence of glucose.

The addition of a few drops of nitrate of bismuth, and then of a solution of potash or soda (Böttger's test) gave a white precipitate, which became black on boiling, in accord with the effect produced by sugar.

Boiling with Fehling's solution gave rise to the production of an immediate and copious precipitate of the reduced oxide of copper. For this, however, the complete removal of the ammonia used in the preparation of the product was necessary. With incomplete removal, the neat and decided reaction otherwise observed was replaced by decoloration and delay of precipitation till after somewhat prolonged boiling. It is to be remarked also that where lime, derived from the employment of animal charcoal which had been imperfectly freed from this principle, was permitted to be present, an interference with the proper reaction of the test was occasioned by the semi-gelatinous precipitate it produced.

The tests which have been referred to can only be said, strictly speaking, to afford presumptive evidence of the existence of sugar. The occurrence of fermentation, however, manifested by the production of alcohol and carbonic acid gas, may be regarded as supplying absolute proof of its presence. On exposing the product derived from the urine to the influence of yeast, I expected, looking at the statements that had been made and the reactions afforded by the other tests, to meet with decided evidence of the occurrence of fermentation, and was surprised to find that it did not ensue. Not only had fermentation been described as taking place, but estimations of the amount of sugar present had been made through the carbonic acid gas evolved. That the fault was not on the part of the yeast I employed was shown by the brisk fermentation which occurred when a portion of it was placed in an aqueous solution of grape sugar and exposed to the suitable temperature. At a loss to account for the negative results that I obtained, I tried the effect of experimenting with urine to which sugar had been purposely added. Again the result was of a negative character. It thus became presumable that there was something connected with the state of the product which checked the activity of the ferment. I noticed that a strongly acid condition existed, and it occurred to me to ascertain whether this might exert an arresting influence on ferment action. I thus was led to neutralise the product with carbonate of soda, and I then found that fermentation actively proceeded. Through not having recognised, therefore, the necessity of giving attention to this point, I failed in my first attempts to obtain fermentation. The circumstances observed stand, it is to be remarked, in accord with the statement of Pasteur, that acidity opposes whilst alkalinity favours the occurrence of fermentation.

In addition to the evidence already brought forward founded upon methods of testing with which we have been long familiar, further evidence is now adducible, derived from the application of other methods which have been recently discovered to constitute valuable tests for sugar.

Phenylhydrazine, already frequently referred to in this work, is an agent which forms crystalline compounds—osazones—with the various sugars. A convenient method of applying the test to healthy urine is the following :—The sugar is thrown down in the manner I have already had occasion to describe (p. 179), by the hydrated oxide of lead. The precipitate containing the lead compound is washed with water (it is not here necessary as for the application of the copper test to wash to such an extent as to get rid of the whole of the ammonia), dissolved in acetic acid to liberate the sugar, and the solution treated with sulphuric acid for the precipitation and separation of the lead. After filtration, the liquid, which already contains the acetic acid wanted, is treated with phenylhydrazine in the requisite proportion, heated on the water-bath for about an hour, and afterwards set aside to cool. In the course of some hours osazone crystals separate out. Subjoined are photo-engravings from microphotographs of osazones from different kinds of urine.

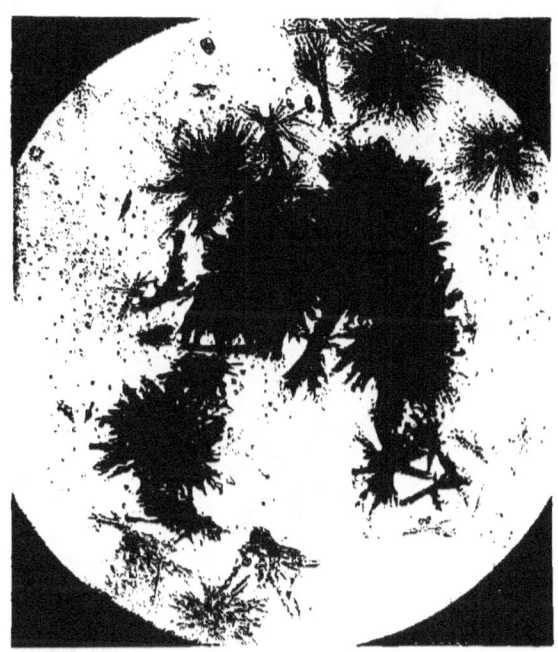

Osazone crystals from the sugar of healthy human urine.
Magnified 400 diameters.

Osazone crystals from the sugar of horse's urine. Magnified 400 diameters.

SUGAR IN HEALTHY URINE.

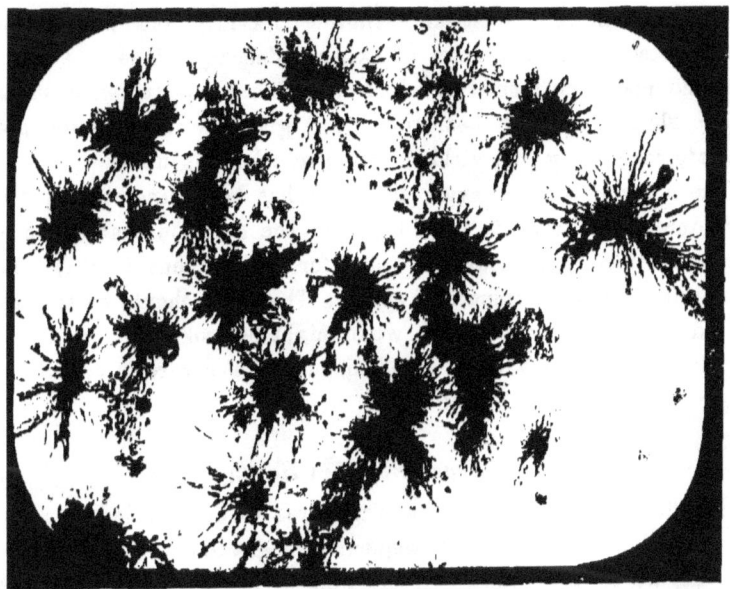

Osazone crystals from the sugar of the urine of the dog fed on animal food. Magnified 400 diameters.

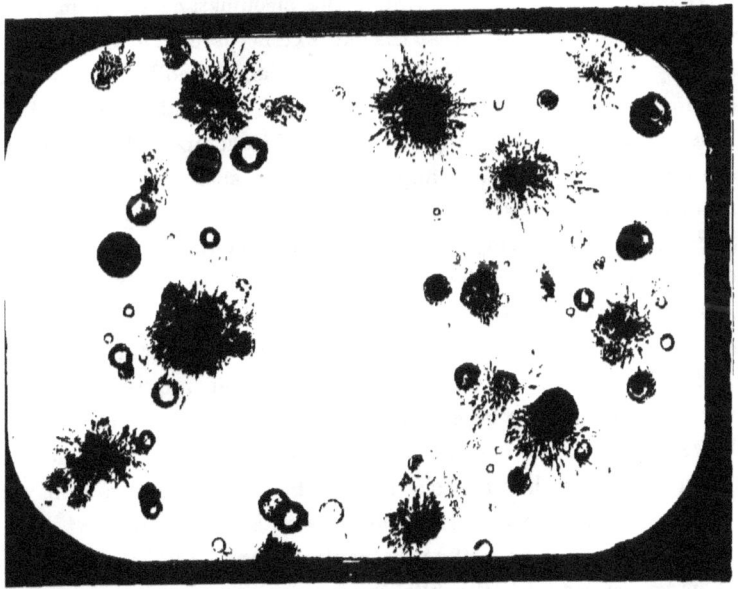

Osazone crystals from the sugar of rabbit's urine. Magnified 400 diameters. The globular masses consist of free phenyl-hydrazine.

Benzoyl chloride constitutes another recently introduced test which possesses the character of extreme delicacy. The reaction is founded upon the formation of an insoluble compound with sugar. When shaken up with healthy urine it gives a white granular precipitate resembling that yielded with a solution of glucose.

The accumulation of evidence which has been presented may be considered, I take it, to absolutely set the question that has been discussed at rest, and leave no room for doubt about healthy urine containing a certain amount of sugar. That such should be the case is, indeed, only in accord with what might be looked for, in view of the fact, about which there is no dispute, that a certain amount of sugar is present in the blood.

Amount of Sugar present in Healthy Urine.

Satisfied upon the point that sugar was to be regarded as a constituent of normal urine, I subsequently sought to obtain information with regard to its amount. At the time of my first giving attention to the subject the ammoniated cupric test had not been introduced and I employed the gravimetric process which has been previously referred to in these pages. Under the circumstances existing, this process is open to the risk that the full extent of cupric oxide reduction may not be represented by the suboxide collected, in consequence of a little ammonia having escaped removal in the washing of the precipitate containing the lead oxide compound. With the ammoniated cupric test, it happens that no such risk of misrepresentation of cupric oxide reduction is incurred, and it altogether affords a far more ready and advantageous method for application. Moreover, we are here placed upon different ground, for the test admits of being directly applied to the urine itself, thus dispensing with the previous separation of the sugar. The uric acid possessing, as it is known to do, cupric oxide reducing power, requires to be first removed, and this may be effected by precipitation with the neutral followed by the basic acetate of lead. Without the employment of the latter some of the uric acid fails to be precipitated, and may be subsequently recognised by crystallisation, contrary to what is the case when the basic as well as the neutral acetate has been employed. The sugar passes with the filtrate, from which the surplus lead is removed by sulphuric acid cautiously added till it no longer precipitates. After filtration potash is used to render alkaline, and titration is then performed with the test solution. If the cupric oxide reduction found to occur be read off as produced entirely by sugar, the amount of this principle ordinarily existing in healthy urine may be said to stand at about 0·5 or a little over per 1000.

Relation of the Sugar of the Urine to that of the Blood.

Under the presumption that a negative behaviour with the ordinary mode of testing was indicative of an absence of sugar, Bernard contended that up to a certain point sugar might exist in the blood without passing off with the urine.

In support of this, he stated that if half a gram of grape sugar per kilo. of body-weight were injected into the jugular vein of a rabbit none appeared in the urine, whilst if a gram per kilo. were employed a certain amount was to be found. He went further, and asserted that in the dog the blood might contain 2·40 per 1000 of sugar without any being discoverable in the urine, whilst if it contained 2·60 per 1000, sugar was to be met with. The mean, or 2·50 per 1000, he spoke of as representing the amount that could be tolerated in the blood without passing off with the urine. With reference to this last statement, it may be safely assumed that his experimental basis was at fault, for, as will be seen later on, 2·50 per 1000 of sugar in the blood constitutes a condition that is attended with the escape of a notable amount of sugar with the urine. In the case of the injection of the half gram and the 1 gram of sugar per kilo. into the circulation of the rabbit, the difference observable resolves itself into a question of whether the amount reaching the urine sufficed to be revealed by the sensitiveness of the test or not.

In reality, up to a certain point, sugar may exist in the urine without being indicated by the ordinary mode of testing, and the difference between urine which does, and that which does not, give a reaction is simply a difference of degree as regards amount, instead of, as it presents the appearance of being, a difference of kind.

In Bernard's experiments sugar was injected into the circulation and the urine subsequently examined with the ordinary copper test, a test which, as I have pointed out, has a limit of sensitiveness that permits a certain small amount of sugar to escape being revealed. With the urine of the rabbit especially, which is often much loaded with solid matter, the development of ammonia in the application of the test may be such as to conceal a considerable amount of sugar. This I have ascertained by comparing the reaction with that given by the ammoniated cupric test. As much as 5, 6, or 7 per 1000, or even more, may be shown to be present by the ammoniated cupric test, without anything but an obscure reaction being given with Fehling's solution.

I have undertaken a set of experiments with the view of penetrating further into the matter under consideration. Sugar was injected into the circulation, and the sugar present in the blood determined, in some instances immediately, and in others at the end of different periods. Where the urine was obtainable, quantitative determinations of the sugar which it contained were also made. The animals experimented upon were rabbits of about 2 kilos. weight. From 15 to 20 c.c. of honey solution, containing definite quantities of glucose (determined by titration), were in each case injected into the jugular vein. In order that the normal state might as nearly as possible prevail, the employment of an anæsthetic was purposely avoided. Steps were taken to secure that the animal remained tranquil during the operative procedure upon the vein, and, as a matter of fact, but little notice seemed to be taken by it of what was done. For obtaining the blood, the animal was killed by pithing, the chest opened, and the right heart incised, the whole being done as rapidly as possible in order to preclude the influence of *post-mortem* change. The results of the experiments stood as follows:—

Injection of 1 gram of Glucose per kilo. of Body Weight.

Rabbit I—
   Killed at once; no urine obtainable.

	Sugar per 1000, expressed as glucose.
Cardiac blood { before sulphuric acid	5·050
after ,, ,,	5·000

Rabbit II—
   Killed at once; no urine obtainable.

Cardiac blood { before sulphuric acid	5·555
after ,, ,,	5·263

Rabbit III—
   Killed 30 minutes after the injection;
   no urine obtainable.

Cardiac blood { before sulphuric acid	2·450
after ,, ,,	2·703

Rabbit IV—
  Killed 1 hour after the injection.

		Sugar per 1000, expressed as glucose.
Cardiac blood	{ before sulphuric acid	1·460
	{ after ,, ,,	1·666
Urine obtained	{ before sulphuric acid	55·160
after death.	{ after ,, ,,	55·160

Rabbit V—
  Killed 1½ hours after the injection.

Cardiac blood	{ before sulphuric acid	1·243
	{ after ,, ,,	1·533
Urine obtained	{ before sulphuric acid	11·370
after death.	{ after ,, ,,	21·150

Injection of 0·5 gram of Glucose per kilo. of Body Weight.

Rabbit VI—
  Killed at once; no urine obtainable.

		Sugar per 1000, expressed as glucose.
Cardiac blood	{ before sulphuric acid	2·380
	{ after ,, ,,	2·548

Rabbit VII—
  Killed 30 minutes after the injection; no urine obtainable.

Cardiac blood	{ before sulphuric acid	1·573
	{ after ,, ,,	1·573

Rabbit VIII—
  Killed 1 hour after the injection.

Cardiac blood	{ before sulphuric acid	1·390
	{ after ,, ,,	1·370
Urine obtained	{ before sulphuric acid	11·370
after death.	{ after ,, ,,	17·860

These experiments fail to support the proposition of Bernard regarding a tolerating capacity of the system for a certain amount of sugar. The results obtained were such as might naturally be looked for with the introduction of a diffusible substance like sugar into the circulation. Where 1 gram per kilo. was injected, and the blood was examined immediately, the amounts discoverable are observed to stand in close conformity. From the amounts found immediately after the injection, both of the gram and the half gram quantities, a steady fall is traceable in those met with at subsequent periods. At the end, however, of an hour and a half the fall had not reached the point representative of the normal condition. The urine, where it was obtainable, showed the presence, even after the injection of the half gram per kilo., of a considerable amount of sugar, and in one of the experiments after the injection of the one gram the quantity amounted to 55 per 1000.

It must not escape notice, in experiments of the nature under consideration, that the state of the bladder, as regards fullness or emptiness at the time of the injection, will form an important factor in determining the character of the result obtained, by influencing the extent to which the urine secreted after the injection is diluted by that previously existing.

A point deserving a passing comment is that in some of the instances a sugar with a lower cupric oxide reducing power than that of glucose was present, notwithstanding glucose was the form of sugar injected.

In another series of experiments, lævulose derived from the recently devised process of manufacture was injected into the circulation in place of the glucose (mixture of dextrose and lævulose) of which honey consists. 1 gram to the kilo. of body weight was the quantity used, and the steps of operative and analytical procedure were precisely the same as in the experiments recorded above.

Injection of 1 gram of Lævulose per kilo. of Body Weight.

```
Rabbit I—                                          Sugar per 1000,
    Killed at once.                              expressed as glucose.
        Cardiac blood  { before sulphuric acid ............  4·387
                       { after     ,,      ,,  ............  3·787
        Urine obtained { before sulphuric acid ............  3·571
        after death .  { after     ,,      ,,  ............  6·250
Rabbit II—
    Killed 30 minutes after the injection.
        Cardiac blood  { before sulphuric acid ............  2·523
                       { after     ,,      ,,  ............  2·577
        Urine obtained { before sulphuric acid ............ 97·402
        after death .  { after     ,,      ,,  ............ 71·091
Rabbit III—
    Killed 1 hour after the injection.
        Cardiac blood  { before sulphuric acid ............  1·893
                       { after     ,,      ,,  ............  1·800
        Urine obtained { before sulphuric acid ............ 37·037
        after death .  { after     ,,      ,,  ............ 35·333
Rabbit IV—
    Killed 1½ hours after the injection.
        Cardiac blood  { before sulphuric acid ............  1·603
                       { after     ,,      ,,  ............  1·517
        Urine obtained { before sulphuric acid ............ 30·303
        after death .  { after     ,,      ,,  ............ 25·900
```

On looking through these results and comparing them with those belonging to the preceding series in which 1 gram per kilo. of glucose from honey was injected, it is noticeable that no significant difference in the amount of sugar found in the blood exists between the two. It fortunately happened that urine was procurable from each of the rabbits after death. In that obtained from rabbit II, the proportion of sugar was strikingly large, whilst with rabbits III and IV a notable amount was also found. The conspicuous point of difference observable from the preceding series is in the fall in the amount of sugar occasioned by boiling with sulphuric acid. This effect of the treatment with acid is only in harmony with the known property of lævulose in becoming more easily destroyed or lost by the action of heat and acids than dextrose. From the urine of rabbit I, in which only the sugar naturally belonging to the healthy state was present, the figures obtained after the treatment with sulphuric acid are higher than those yielded before the treatment. The inverted condenser, it may be remarked, and not the autoclave, was employed in the case of all the urines.

As the outcome of all that has preceded, not only may it be said that sugar normally exists in the urine, but further that it is present in proportion to the amount contained in the blood; and this holds good for the slight fluctuations occurring under ordinary conditions as well as for the larger amounts occurring in connection with diabetes.

In illustration of this last assertion, I may give the results of some observations which I conducted some years ago upon persons suffering from diabetes, and which were recorded in the Croonian Lectures, "On certain points connected with Diabetes," 1878, p. 80. Blood obtained by cupping was analysed, and the urine passed during the corresponding twenty-four hours' period was collected, measured, and examined.

In the appended table the particulars are arranged so as to show the salient points belonging to each observation. Case I was that of a patient suffering from a severe form of diabetes, and subsisting at the time upon a diet of ordinary mixed food. II and III were also severe cases, and in these, observations were conducted when the patients were upon an ordinary mixed diet, and again when the sugar to be eliminated had been diminished by the exclusion of starchy and saccharine matter from the food. Case IV was one of a milder type, and the sugar passed under a partially restricted diet was less than that passed under the restricted diet in the other cases.

Nothing could show more clearly than these results that the amount of sugar appearing in the urine stands in proportion to the amount existing in the blood. The urine in reality constitutes the channel of exit for the sugar which is abnormally present in diabetes, and in this way affords the natural provision for keeping down the amount within the system. Such being what is noticeable with reference to the sugar belonging to diabetes, it would indeed be strange if a different principle prevailed with reference to that belonging to health, and if, with a diffusible substance like sugar, the contention of Bernard held good that a hard and fast line existed between escape and non-escape with the urine. The fact, indeed, is that a small amount of sugar, as has been shown, exists in healthy urine, in harmony with the small amount existing in healthy blood. As the amount rises in the blood so it rises in the urine, without any absolutely sharp line of demarcation existing between the normal and abnormal states. The transition, in fact, from health to disease, instead of being abrupt, takes place by gradations of an insensible nature.

Comparative State of Blood and Urine in Diabetes.

	Quantity per 24 hours.	Urine. Specific gravity.	Sugar per 24 hours.	Sugar per 1000 parts.	Blood. Sugar per 1000 parts (mean of 2 analyses).
CASE I. January 5th. Mixed diet............	6608 c.c.	1040	751·6 grams	109·91	5·763
CASE II. January 8th. Mixed diet............ January 28th. Restricted diet.........	6474 ,, 3407 ,,	1041 1031	633·0 ,, 245·2 ,,	94·08 61·34	5·545 2·625
CASE III. June 8th. Mixed diet............ July 20th. Restricted diet.........	5878 ,, 2470 ,,	1036 1033	567·7 ,, 115·8 ,,	93·39 45·49	4·970 2·789
CASE IV. March 9th. Partially restricted diet...... June 28th. Partially restricted diet......	1704 ,, 852 ,,	1036 1034	84·1 ,, 27·9 ,,	48·11 31·76	1·848 1·543

MUSCLE IN RELATION TO SUGAR.

In accord with what is noticeable elsewhere throughout the system, muscle contains a certain amount of sugar. I have made a large number of observations upon the sugar belonging to it, and will here furnish a representation of the information which has been obtained.

Mammalian Animals.—The following digest gives the main points of information derived from 115 determinations of muscle-sugar in the dog, rabbit, cat, sheep, horse, and pig.

Nature of Sugar.

As regards the nature of the sugar met with, it is to be stated that the cupric oxide reducing power belonging to it stood, in the great majority of instances, below that of glucose. The ordinary range of reducing power was from about 45 or 50 to about 80, as compared with that of glucose at 100. The average was found to be about 65.

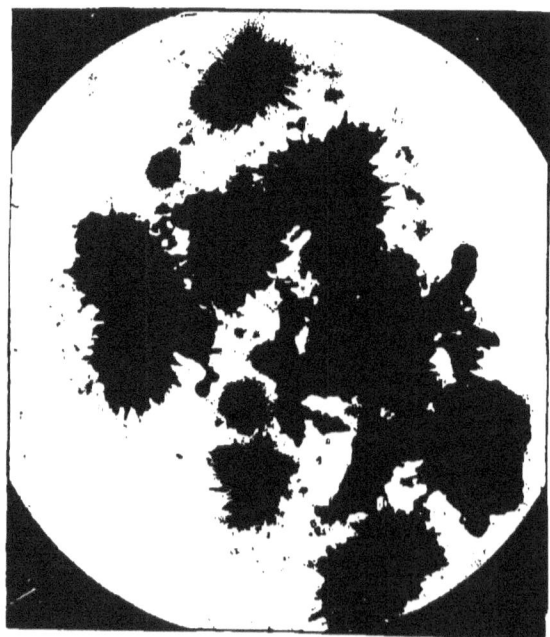

Osazone crystals from the sugar of muscle (bullock).
Magnified 400 diameters.

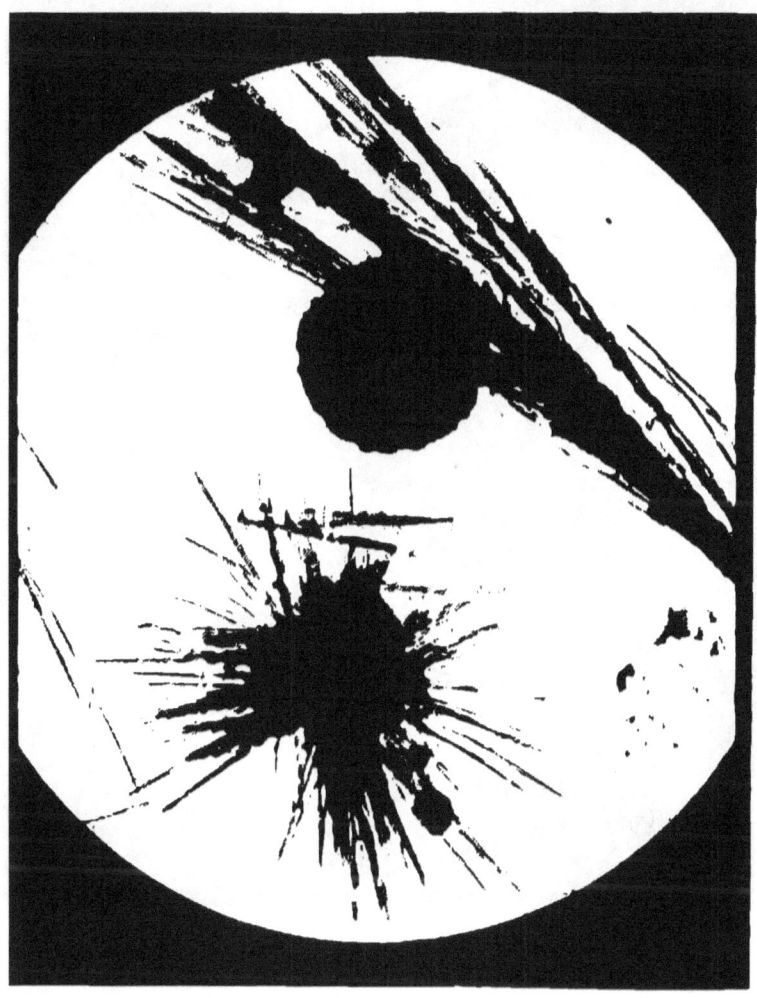

Osazone crystals from the sugar of leg muscle (dog). Magnified 400 diameters.

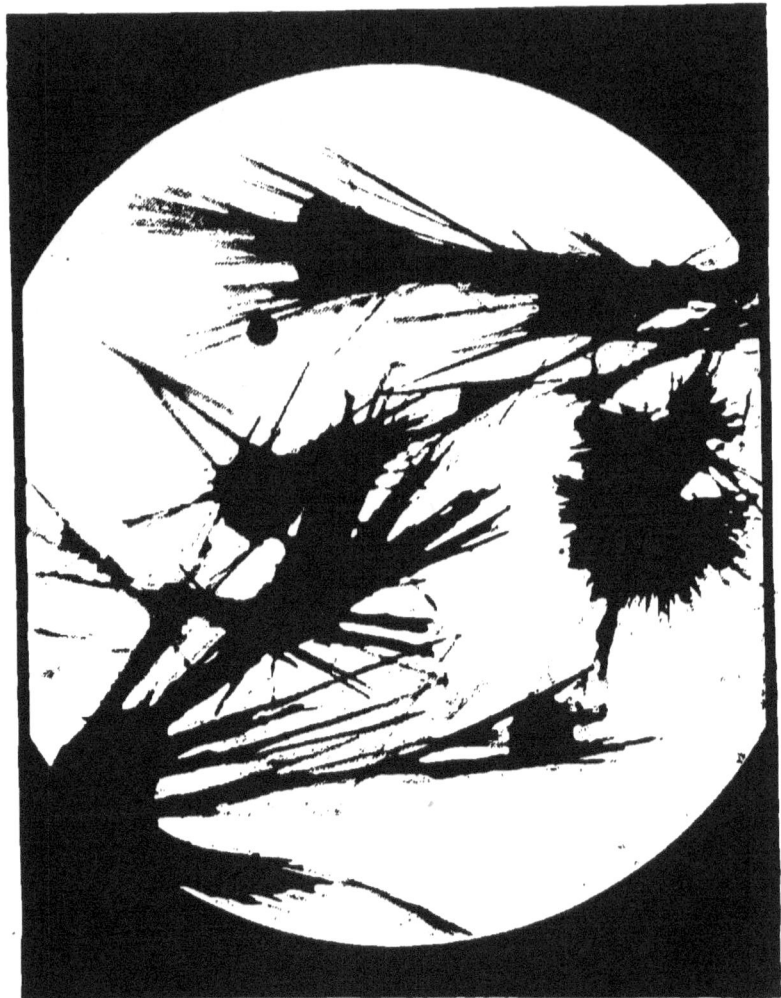

Osazone crystals from the sugar of cardiac muscle (dog). Magnified 400 diameters.

Amount of Sugar.

The amount of sugar, estimated in the form of glucose, stood in the largest number of instances at from 2 to 4 per 1000. In no instance was the quantity found to be less than 1 per 1000. Rarely was it observed to exceed 5 or 6 per 1000. Upon one occasion, however, as much as 9 per 1000 was found. The instance falls in the series of analyses given a little further on of muscular tissue taken from different parts of a dog, and applies to the muscle of the hind leg. The accompanying glycogen figures stood, it was likewise noticeable, unusually high. Another large amount that I find recorded was yielded by the muscular tissue of the diaphragm of a dog, the figures for which stood at 3·800 per 1000 before treatment with sulphuric acid, and 7·100 per 1000 afterwards. In this case, as in the other, muscular tissue taken from different parts was examined. The analytical figures for the other parts were of an ordinary character. The amount of the accompanying glycogen was below the usual.

To ascertain whether an ordinarily conducted examination furnishes a correct representation of the amount of muscle-sugar existing during life, a piece of muscle was in several instances removed quickly after death, and placed in a freezing mixture, whilst another piece was taken in an ordinary way. No difference of a definite nature was encountered. The following detailed account of an observation may be given :—

Rabbit—		Sugar per 1000, expressed as glucose.	Cupric oxide reducing power of the sugar present in relation to that of glucose at 100.
Muscle of fore-leg—			
Frozen portion	before sulphuric acid	1·747	} 64
	after ,, ,,	2·730	
Unfrozen portion	before ,, ,,	1·983	} 77
	after ,, ,,	2·567	
Muscle of hind-leg—			
Frozen portion	before ,, ,,	2·130	} 59
	after ,, ,,	3·640	
Unfrozen portion	before ,, ,,	2·427	} 67
	after ,, ,,	3·640	

Amount of Sugar in Muscular Tissue derived from different parts.

The muscular tissue taken from different parts of the same animal was sometimes found to present considerable variation in relation to the amount of sugar present. The following is an example in illustration :—

		Sugar per 1000, expressed as glucose.	Cupric oxide reducing power of the sugar present in relation to that of glucose at 100.
Dog—			
Muscle of tongue	before sulphuric acid	2·458	} 66
	after ,, ,,	3·727	
,, diaphragm	before ,, ,,	1·662	} 74
	after ,, ,,	2·259	
,, heart	before ,, ,,	2·855	} 61
	after ,, ,,	4·700	
,, hind-leg	before ,, ,,	6·502	} 70
	after ,, ,,	9·292	

Nothing was deducible from the observations pointing to the existence of any specific difference associated with a difference in the kind of animal.

Question of the Influence of Food.

No effect was traceable to food, except that in a group of instances where sugar was administered in conjunction with other food a rather higher range of figures was presented.

Non-Mammalian Animals :—I have given a summarised, instead of a detailed, representation of the observations conducted upon the mammalian animal, on account of the large number to be dealt with. Those upon other animals, being comparatively few, may be given in full. Two points are noticeable in connexion with them. The amount of sugar is in general less, and its cupric oxide reducing power, especially in the case of the crustacean animal, lower—in some instances indeed conspicuously lower—than in the mammal.

		Sugar per 1000, expressed as glucose.	Cupric oxide reducing power of the sugar present in relation to that of glucose at 100.
Fowl—			
Muscle of breast	before sulphuric acid	1·380	} 59
	after ,, ,,	2·340	
,, leg	before ,, ,,	1·120	} 58
	after ,, ,,	1·930	
Grouse—			
Muscle of breast	before ,, ,,	1·130	} 41
	after ,, ,,	2·730	

					Sugar per 1000, expressed as glucose.	Cupric oxide reducing power of the sugar present in relation to that of glucose at 100.
Tortoise—						
Muscle	before sulphuric acid....				1·220	} 67
	after	,,	,,	....	1·810	
Tortoise—						
Muscle	before	,,	,,	....	1·272	} 56
	after	,,	,,	....	2·254	
Frog—						
Muscle	before	,,	,,	....	0·970	} 32
	after	,,	,,	....	3·030	
Frog—						
Muscle	before	,,	,,	....	0·975	} 42
	after	,,	,,	....	2·330	
Salmon—						
Muscle	before	,,	,,	....	0·530	} 23
	after	,,	,,	....	2·300	
Cod—						
Muscle	before	,,	,,	....	0·830	} 38
	after	,,	,,	....	2·170	
Turbot—						
Muscle	before	,,	,,	....	0·830	} 57
	after	,,	,,	....	1·460	
Mackerel—						
Muscle	before	,,	,,	....	2·469	} 55
	after	,,	,,	....	4·502	
Plaice—						
Muscle	before	,,	,,	....	2·453	} 75
	after	,,	,,	....	3·240	
Eel—						
Muscle	before	,,	,,	....	0·800	} 39
	after	,,	,,	....	2·030	
Muscle .... (another portion)	before	,,	,,	....	0·700	} 38
	after	,,	,,	....	1·810	
Lobster—						
Muscle of tail..	before	,,	,,	....	1·037	} 60
	after	,,	,,	....	1·720	
,, claw	before	,,	,,	....	0·780	} 28
	after	,,	,,	....	2·730	
Lobster—						
Muscle of tail..	before	,,	,,	....	0·620	} 37
	after	,,	,,	....	1·660	
,, claw	before	,,	,,	....	1·760	} 83
	after	,,	,,	....	2·090	
Lobster—						
Muscle of tail..	before	,,	,,	....	0·230	} 31
	after	,,	,,	....	0·743	
,, claw	before	,,	,,	....	0·270	} 43
	after	,,	,,	....	0·624	
Crab—						
Muscle of claw	before	,,	,,	....	1·500	} 100
	after	,,	,,	....	1·500	
Crab—						
Muscle of claw	before	,,	,,	....	1·482	} 62
	after	,,	,,	....	2·392	

THE SPLEEN IN RELATION TO SUGAR.

Representative examples from a collection of twenty observations.

		Sugar per 1000, expressed as glucose.	Cupric oxide reducing power of the sugar present in relation to that of glucose at 100.
Spleen of horse ..	{ before sulphuric acid.... { after ,, ,, ...	1·320 1·634	} 81
Spleen of sheep ..	{ before { after	1·984 2·777	} 71
Spleen of dog	{ before ,, ,, ... { after ,, ,, ...	1·587 1·827	} 87
Spleen of dog	{ before ,, ,, ... { after ,, ,, ...	2·032 2·777	} 73
Spleen of dog	{ before ,, ,, ... { after ,, ...	3·342 5·819	} 57

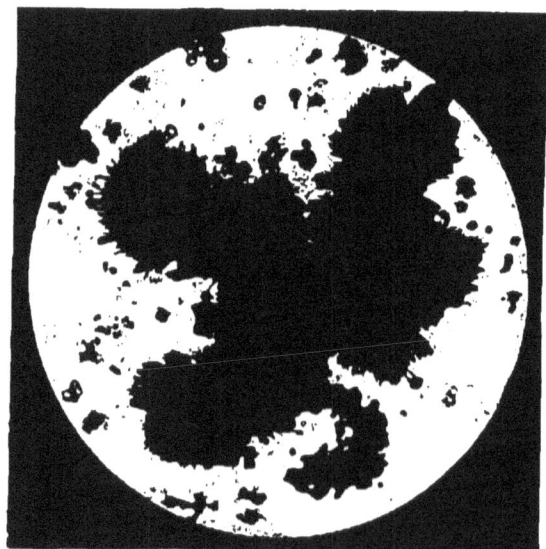

Osazone crystals from the sugar of the spleen (horse).
Magnified 400 diameters.

THE KIDNEY IN RELATION TO SUGAR.

Representative examples from a collection of fifteen observations.

		Sugar per 1000, expressed as glucose.	Cupric oxide reducing power of the sugar present in relation to that of glucose at 100.
Kidney of horse..	{ before sulphuric acid.... { after ,, ,, ...	1·572 2·575	} 61
Kidney of horse..	{ before ,, ,, ... { after ,, ,, ...	0·896 1·132	} 79
Kidney of dog ...	{ before ,, ,, ... { after ,, ,, ...	1·685 2·229	} 76
Kidney of dog ...	{ before ,, ,, ... { after ,, ,, ...	1·123 1·765	} 64
Kidney of sheep..	{ before ,, ,, ... { after ,, ,, ...	0·960 1·380	} 70

Osazone crystals from the sugar of the kidney (horse). Magnified 400 diameters.

THE PANCREAS IN RELATION TO SUGAR.

Representative examples from a collection of eighteen observations.

		Sugar per 1000, expressed as glucose.	Cupric oxide reducing power of the sugar present in relation to that of glucose at 100.
Pancreas of horse	{ before sulphuric acid.... { after ,, ,, ...	1·463 2·688	} 54
Pancreas of sheep	{ before ,, ,, ... { after ,, ,, ...	0·731 1·388	} 53
Pancreas of sheep	{ before ,, ,, ... { after ,, ,, ...	0·712 0·725	} 98
Pancreas of dog..	{ before ,, ,, ... { after ,, ,, ...	0·812 1·433	} 57
Pancreas of dog..	{ before ,, ,, ... { after ,, ,, ...	2·314 3·343	} 69

THE LUNG IN RELATION TO SUGAR.

Four observations.

		Sugar per 1000, expressed as glucose.	Cupric oxide reducing power of the sugar present in relation to that of glucose at 100.
Lungs of horse ..	{ before sulphuric acid.... { after ,, ,, ...	1·545 2·183	} 71
Lungs of dog	{ before ,, ,, ... { after ,, ,, ...	3·186 4·870	} 65
Lungs of dog	{ before ,, ,, ... { after ,, ,, ...	0·947 1·406	} 67
Lungs of fœtal pups	{ before ,, ,, ... { after ,, ,, ...	1·631 1·725	} 95

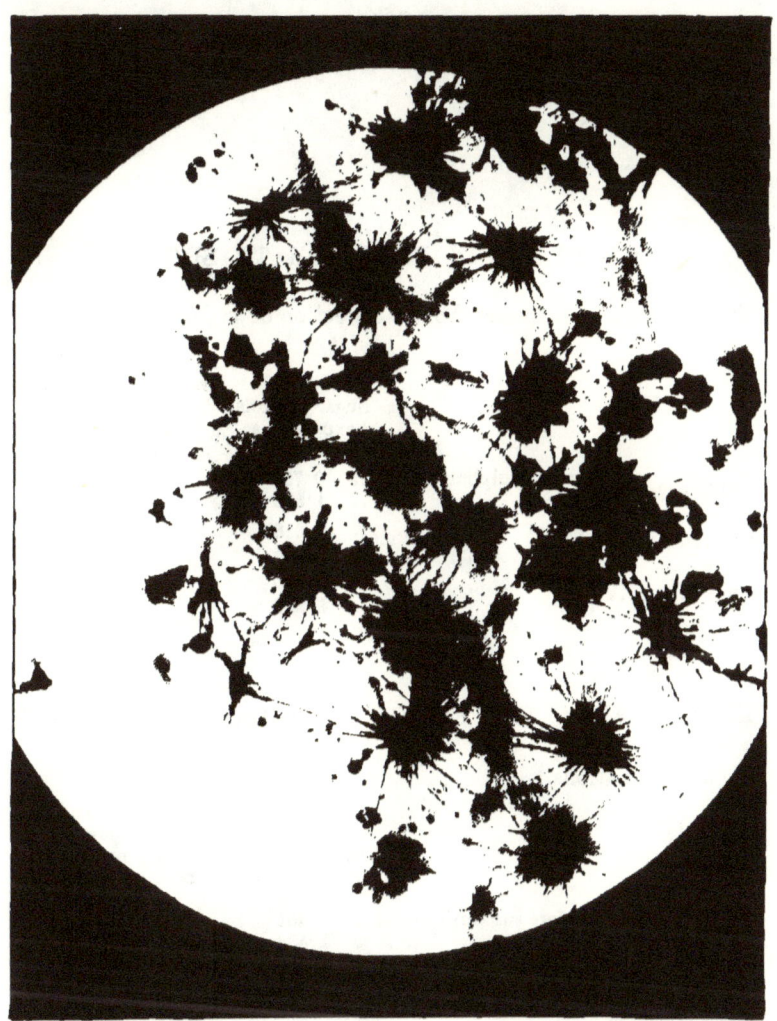

Osazone crystals from the sugar of the lung (dog). Magnified 100 diameters.

THE BRAIN IN RELATION TO SUGAR.

Representative examples from a collection of six observations.

		Sugar per 1000, expressed as glucose.	Cupric oxide reducing power of the sugar present in relation to that of glucose at 100.
Brain of dog.....	{ before sulphuric acid.... { after ,, ,, ...	0·780 1·500	} 52
Brain of dog.....	{ before ,, ,, ... { after ,, ,, ...	1·318 1·705	} 77
Brain of dog.....	{ before ,, ,, ... { after ,, ,, ...	0·720 1·216	} 59

THE PLACENTA AND FŒTUS IN RELATION TO SUGAR.

Four observations.

		Sugar per 1000, expressed as glucose.	Cupric oxide reducing power of the sugar present in relation to that of glucose at 100.
Placenta of dog..	{ before sulphuric acid.... { after ,, ,, ...	1·461 1·438	} glucose
Placenta of rabbit	{ before ,, ,, ... { after ,, ,, ...	1·263 2·285	} 55
Fœtuses removed from the uterus of a recently killed rabbit	{ before ,, ,, ... { after ,, ,, ...	2·845 4·699	} 61
Placenta and early fœtuses of rabbit	{ before ,, ,, ... { after ,, ,, ...	1·411 2·043	} 69

ORGANS OF GENERATION OF FISH AND CRUSTACEA IN RELATION TO SUGAR.

		Sugar per 1000, expressed as glucose.	Cupric oxide reducing power of the sugar present in relation to that of glucose at 100.
Cod—			
Spermatic organ	before sulphuric acid....	0·430	} 22
	after ,, ,, ...	1·960	
Ovary	before ,, ,, ...	0·630	} 32
	after ,, ,, ...	1·950	
Mackerel—			
Spermatic organ	before ,, ,, ...	2·300	} 62
	after ,, ,, ...	3·680	
Ovary	before ,, ,, ...	1·530	} 37
	after ,, ,, ...	4·090	
Lobster—			
Ovary	before ,, ,, ...	3·440	} 47
	after ,, ,, ...	7·380	
Lobster—			
Ovary	before ,, ,, ...	3·600	} 100
	after ,, ,, ...	3·600	
Ova attached to tail	before ,, ,, ...	1·430	} 55
	after ,, ,, ...	2·570	
Crab—			
Ovary	before ,, ,, ...	2·600	} 60
	after ,, ,, ...	4·300	
Crab—			
Spermatic organ	before ,, ,, ...	2·020	} 42
	after ,, ,, ...	4·750	

THE EGG IN RELATION TO SUGAR.

The egg agrees with other animal products in containing a certain amount of sugar. My observations have been conducted upon the egg of the domestic fowl, and I have carried my enquiry to the condition existing during the progress of incubation.

Taking the entire egg and including the shell in the initial weight, the following figures show the amount and nature of the sugar found in three analyses:—

		Sugar per 1000, expressed as glucose.	Cupric oxide reducing power of the sugar present in relation to that of glucose at 100.
Fresh egg—			
1.	before sulphuric acid.... after ,, ,, ...	1·962 1·962	} glucose
2.	before ,, ,, ... after ,, ,, ...	2·560 2·560	} glucose
3.	before ,, ,, ... after ,, ,, ...	2·554 2·554	} glucose

An examination of the white and yolk separately gave the subjoined results. The egg taken weighed, including the shell, 63·13 grams. The white obtained from it weighed 37·06 grams, and the yolk 18·75 grams. The sugar figures respectively yielded stood as follows:—

		Sugar per 1000, expressed as glucose.	Cupric oxide reducing power of the sugar present in relation to that of glucose at 100.
White	before sulphuric acid.... after ,, ,, ...	3·418 3·262	} glucose
Yolk	before ,, ,, ... after ,, ,, ...	0·789 0·832	} 95

Nearly mature ova, taken from the ovary and therefore comprehending only the yolk, gave, on examination, the following figures:—

		Sugar per 1000, expressed as glucose.	Cupric oxide reducing power of the sugar present in relation to that of glucose at 100.
Nearly mature ova taken from the ovary	before sulphuric acid.... after ,, ,, ...	0·805 0·808	} glucose

It thus appears that the kind of sugar met with is, in both white and yolk, shown to be practically glucose, and that the white contains a far larger amount than the yolk. For the entire egg the quantity stands at from about 2 to about 2·5 per 1000.

The character of the osazone derivable from the sugar of the egg confirms what has just been stated regarding the form of sugar present. Subjoined is a photo-engraving from a micro-photograph of the crystals obtained from an aqueous extract of the white of egg treated with phenylhydrazine. They constitute typical crystals of glucosazone.

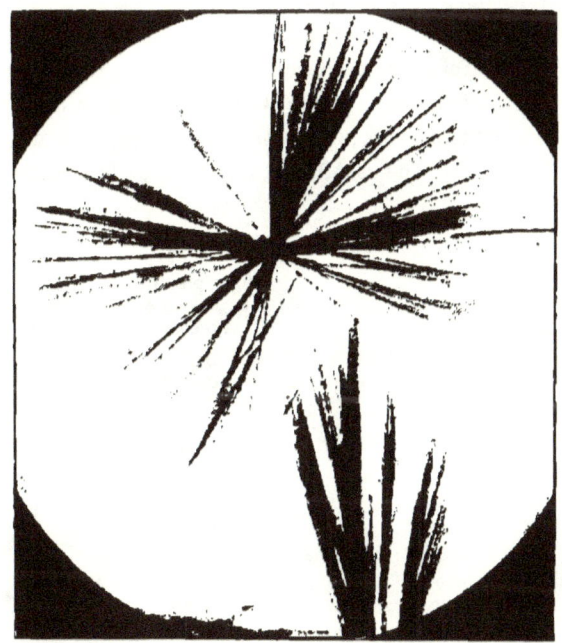

Osazone crystals from the sugar of the egg. Magnified 400 diameters.

The effect of incubation was studied through the employment of an incubator. Pains were taken to procure fertile eggs, but many of them, nevertheless, failed to undergo any development. In others development proceeded up to a certain point and then stopped. It further unfortunately happened, in connexion with a batch, that just before the completion of the full term of incubation the temperature was permitted to fall below the proper point, which led to several nearly mature chicks being found dead within the shell. It was found in all cases that, under exposure in the incubator, a loss of weight steadily occurred, amounting, at the end of the twenty-one days term of incubation, to something considerable. In one instance it was noticed to have reached to as much as over 20 grams. To secure a uniform basis of comparison, the initial weight of the egg was taken in making the calculation. If calculated in relation to the diminished weight existing, higher proportionate figures would be presented, giving a semblance of the occurrence of an increase in the amount of sugar where no increase had actually taken place, or concealing a fall where such in reality had occurred.

When the egg was removed from the incubator, the shell was broken and a careful examination made, to see if any development had occurred. Of the eggs which were found not to have undergone development, several were submitted to analysis, as well as those which had done so. The results obtained ran as follows:—

Period when taken and condition.		Sugar per 1000, expressed as glucose.	Cupric oxide reducing power of the sugar present in relation to that of glucose at 100.
8th day—			
Live chick	before sulphuric acid....	0·918	} 95
	after ,, ,, ...	0·970	
9th day—			
Live chick	before ,, ,, ...	0·594	} glucose.
	after ,, ,, ...	0·594	
11th day—			
Live chick	before ,, ,, ...	0·733	} 93
	after ,, ,, ...	0·789	
12th day—			
Live feathered chick	before ,, ,, ...	0·637	} 74
	after ,, ,, ...	0·859	
14th day—			
Live feathered chick	before ,, ,, ...	0·618	} 90
	after ,, ,, ...	0·690	
Small dead chick	before ,, ,, ...	0·421	} 98
	after ,, ,, ...	0·427	

INFLUENCE OF INCUBATION.

Period when taken and condition.		Sugar per 1,000, expressed as glucose.	Cupric oxide reducing power of the sugar present in relation to that of glucose at 100.
16th day— Live feathered chick	before sulphuric acid.... after ,, ,, ...	0·578 0·662	} 88
Evidence of some incipient development, but no distinct embryo noticed	before ,, ,, ... after ,, ,, ...	1·191 1·253	} 94
19th day— Small dead chick	before ,, ,, ... after ,, ,, ...	0·514 0·562	} 92
21st day— Fair-sized dead chick of about the 10th day	before ,, ,, ... after ,, ,, ...	0·391 0·474	} 83
20th day— Nearly full-sized dead chick	before ,, ,, ... after ,, ,, ...	0·744 1·068	} 70
21st day— Nearly full-sized dead chick	before ,, ,, ... after ,, ,, ...	0·539 0·688	} 78
21st day— Nearly full-sized dead chick	before ,, ,, ... after ,, ,, ...	0·591 0·827	} 71
8th day— Undeveloped ..	before ,, ,, ... after ,, ,, ...	2·485 2·528	} 98
10th day— Undeveloped ..	before ,, ,, ... after ,, ,, ...	2·797 2·811	} 99
16th day— Undeveloped ..	before ,, ,, ... after ,, ,, ...	2·450 2·206	} glucose.
17th day— No appearance of development	before ,, ,, ... after ,, ,, ...	0·543 0·543	} glucose.
19th day— No appearance of development	before ,, ,, ... after. ,, ,, ...	0·656 0·635	} glucose.
21st day— Undeveloped ..	before ,, ,, ... after ,, ,, ...	1·014 1·218	} 84

On looking at the above results, two points are noticeable—one, a marked and speedy drop in the amount of sugar at the commencement of embryonic development, and the other, an alteration in the character of the sugar, represented by a diminution of cupric oxide reducing power. Two instances, it is observable, occur amongst the undeveloped eggs, in which there was a similar drop in the amount of sugar found, and the question suggests itself, on account of the general accordance in the results, whether there might not here have been some incipient development which had escaped observation.

The analyses, it may be stated, also included the determination, by the process adopted for the estimation of glycogen, of the carbohydrate matter contained in the residue from alcoholic extraction. No alteration was discernible that could be considered attributable to the influence of development. The same, it may be remarked, was found to be the case with the figures expressive of the total carbohydrate encountered.

GLYCOGEN AND PROTEID-CLEAVAGE CARBOHYDRATE.

I have now to speak of the carbohydrate matter, other than sugar, obtainable from the constituent parts of the animal system. After the removal of sugar by alcoholic extraction, the residue still gives evidence of the presence of carbohydrate. With certain tissues, this carbohydrate consists, undoubtedly, to a greater or less extent of glycogen—a principle the discovery of which resulted from the masterly researches of Bernard. In addition, as I have shown in an earlier part of this work when treating of the glucoside constitution of proteid matter, the residue yields, under the process of analysis which I have been in the habit of employing for the separation and estimation of glycogen—namely, boiling with potash and precipitating with alcohol—a certain amount of carbohydrate material, not identical with, but closely allied to, glycogen.

As regards glycogen, it is well-known history that Bernard, in the first instance, recognised it in the liver. Subsequent investigation led to his recognition of it in muscle and in the lungs, whilst in no other tissue of the adult animal did he meet with it. In the prosecution of further research he detected it in cells belonging to the amnion and placenta, and subsequently found it widely distributed through the tissues of the fœtus. From observing its presence in the skin and its epithelial appendages, in the lining surface of the alimentary canal, in the ducts of glands opening into the alimentary canal (not the glands themselves), in the respiratory surface, and in the genito-urinary surface, of the embryo, he was led to regard it as being concerned in the development of the limiting membranes. In the liver he noted that glycogen did not make its appearance until about the middle of intra-uterine life. In fœtal structures other than those mentioned, with the exception of muscle, he did not detect its presence.

In my own work in connexion with the liver, I, at an early period, had recourse for the extraction and estimation of glycogen to a process which consisted, as I have previously mentioned, in boiling with potash, precipitating with alcohol, boiling with dilute sulphuric acid, and titrating with the ammoniated cupric test solution. Circumstances afterwards led to my applying the process to other constituents of the body. I had happened to notice in the case of blood that the residue from alcoholic extraction, on being treated with boiling water, yielded a product possessing a certain small amount of cupric oxide reducing power before treatment with sulphuric acid and an amount considerably greater afterwards. This, after being put to the test of enquiry, I concluded was due to the presence of glycogen together with a little sugar which had escaped removal by the alcohol. I was led now to apply the potash process made use of for the separation of glycogen from the liver to the residue from the alcoholic extraction of blood, and I found that it yielded a non-reducing substance precipitable by alcohol and convertible through the agency of sulphuric acid into a reducing sugar. I afterwards applied the process to the tissues and organs of the body generally, with the result that in each case a similar product was obtained.

Unaware of what I have recently discovered with regard to the glucoside constitution of proteid matter, I looked upon the product obtained as consisting of glycogen. Research, however, has taught me that the proteid entering into the constitution of the blood and other component parts of the body yields, under the potash process employed, a cleavage carbohydrate, which accompanies the glycogen where such exists, and in the cases where no glycogen exists stands alone as the source of the cupric oxide reducing product obtained.

I may mention here that in my former writings I have spoken of glycogen (and, of course, with it was included the cleavage carbohydrate from proteid) under the term "amyloid substance." As I have previously stated, I look upon the term "glycogen" as a misnomer, on account of the unsound physiological basis upon which its application is founded. I have felt that the term "amyloid substance" is one that simply expresses the nature of the material and involves no physiological error, and, therefore, that its adoption in the place of "glycogen" would be exceedingly desirable. The term "glycogen," however, has become so deeply rooted in chemical and physiological literature, that to avoid complication I have considered it wise to sink my own view on the matter, and have employed it throughout in this work, instead of the term "amyloid substance" which I formerly used.

A new difficulty is created by the discovery of the cleavage carbohydrate from proteid matter, as this passes with the glycogen in the results yielded by analysis. I do not at present see how they are satisfactorily to be separately estimated. The results that follow comprise the figures derived from both materials. When the analyses were undertaken I was unaware of what has since been learnt about the increased cupric oxide reducing power which results from the employment of 10 per cent. as compared with 2 per cent. sulphuric acid when the cleavage carbohydrate is being dealt with. As I have shown (p. 37), the increase is to the extent of nearly doubling the cupric oxide reducing power. It follows, therefore, that in the results given, since 2 per cent. was the strength of acid in all cases employed, the cleavage carbohydrate is considerably under-estimated. Moreover, a further source of under-estimation, it must be stated, exists in connexion with the treatment with potash, for I have found that if blood or other proteid-containing matter be mixed with potash and allowed to remain in contact with it for several days before being boiled, a larger amount of cleavage carbohydrate is obtained than if the boiling is performed at once. The results given are derived from analyses in which the boiling (with 10 per cent. potash) was conducted without any previous prolonged standing, and therefore represent an amount of material short of what might have been attainably obtained. They must, therefore, be only taken for what they are really worth—that is, not as affording full or precise information of the state existing in the individual cases, but as giving an approximate representation of the truth relating thereto, and with this a correct representation of the comparative state of the different specimens examined. It will be understood that the remarks which have just been made do not apply to the estimation of glycogen. Whatever of this principle may be present, is fully and accurately represented by the process employed.

In the results that follow, then, the figures referring to the cleavage carbohydrate are considerably under-stated. The cleavage carbohydrate and glycogen I am obliged to express together, as they are both yielded by the process adopted. For the sake of convenience they will be comprised, assuming that the cleavage carbohydrate, like glycogen, is of an amylose nature, under the term "amylose carbohydrate," which will thus represent the carbohydrate obtained from the residue from which the sugar has been removed by alcoholic extraction. How much may be actually cleavage carbohydrate, or how much glycogen, it is impossible definitely to say. All I can venture upon stating is that, in the results that follow, anything of account beyond about 2 or 3, or possibly 4, per 1000 may presumably be regarded as due to glycogen.

Under the description of the analytical steps of procedure given in a former part of this work, I referred (p. 63) to the fallacy that is liable to be occasioned by the employment of paper filters, through contamination of the product with cellulose. Glass-wool was invariably employed as a filter medium in the analyses from which all the results that follow were derived.

BLOOD AND VARIOUS STRUCTURES OF THE BODY IN RELATION TO AMYLOSE CARBOHYDRATE (PROTEID-CLEAVAGE CARBOHYDRATE AND GLYCOGEN).

BLOOD.

Amongst a collection of observations taken under a variety of circumstances and conducted at various times, I find over a hundred which may be selected as representative of ordinary or natural conditions. In these the average figures yielded for the blood of the general circulation stand at about 1 per 1000 or a little over, expressed as glucose, the ordinary limits being from about 0·8 to 1·3.

As far as the results before me show, no evidence is afforded of any appreciable influence being noticeable in the blood of the general circulation as the effect of the ingestion of food. In the case, however, of the portal blood, a decided influence is perceptible, which presents itself in its most pronounced form after the ingestion of starchy food. Under these latter circumstances, the figures met with have frequently stood as high as 2 to 2·5 per 1000, and in one case they reached 2·7 per 1000. In a series of fifty-two observations upon animals taken at a period of fasting, after animal food, and after starchy food, the average figures for the portal and the cardiac blood respectively stood as follows:—

	Amylose carbohydrate per 1000, expressed as glucose.
At a period of fasting—	
Portal blood	1·13
Cardiac blood	1·12
After animal food—	
Portal blood	1·35
Cardiac blood	1·08
After starchy food—	
Portal blood	1·79
Cardiac blood	1·07

Sugar introduced by injection into the general circulation appears to lead to an increase of amylose carbohydrate in the blood. In some experiments upon rabbits where glucose from honey was injected into the jugular vein, not only was a large amount of sugar afterwards found, as shown at p. 189, but also, in most of the instances, an unexpectedly large quantity of amylose carbohydrate. The figures for the latter stood as follows in the several instances :—

	Amylose carbohydrate, expressed as glucose, per 1000 of blood collected from the heart.			
	Rabbits killed at once.	Rabbits killed after ½ hour.	Rabbits killed after 1 hour.	Rabbit killed after 1½ hours.
Three rabbits— ¼ gram of glucose per kilo. of body-weight injected into jugular vein	2·26	2·04	2·20	—
Five rabbits— 1 gram of glucose per kilo. of body-weight injected into jugular vein	1·25 1·25	2·38	1·51	2·02

In another set of experiments on rabbits, in which 1 gram of lævulose per kilo. of body-weight was injected into the jugular vein, the figures obtained stood as follows:—Killed at once, 1·11; after ½ hour, 1·54; after 1 hour, 1·67; and after 1½ hours, 1·67 per 1000.

Muscle.

	Amylose carbohydrate per 1000, expressed as glucose.		
	External muscles.	Heart.	Diaphragm.
Dog.—			
1 {right leg .... / left leg.............	1·58 / 1·53	1·38	2·03
2 {right leg .... / left leg.............	1·25 / 1·13	1·67	1·80
3	16·67	7·72	29·59
4	3·14	1·96	19·61
5	2·34	0·42	—
6	1·23	0·19	—
7	0·95	1·07	—
8	15·67	—	—
9 (fœtal pups)	11·37	1·64	—

MUSCLE. 217

	Amylose carbohydrate per 1000, expressed as glucose.		
	External muscles.	Heart.	Diaphragm.
Horse.—			
1	15·15	7·55	—
2	17·17	7·31	—
3	6·70	6·67	—
4	5·47	4·50	—
5	11·11	2·18	—
6	16·33	—	—
Sheep.—			
1	2·14	1·13	—
2	2·69	1·26	—
3	2·19	1·42	—
4	1·49	2·15	—
Pig.—			
1	3·22	1·54	—
2	3·26	1·21	—
Rabbit.—			
1	0·94	0·66	—
2 {fore leg / hind leg}	0·81 / 1·09	—	—
3 {fore leg / hind leg}	0·54 / 0·54	—	—

In several observations upon muscle taken indiscriminately from different rabbits, the figures obtained stood as follows:—3·13—1·97—1·37—1·25—1·04—1·00 and 0·67 per 1000.

Cat.—In eleven observations upon muscle taken indiscriminately, the figures respectively obtained were—14·62—9·24—7·35—6·80—6·79—5·72—5·93—3·95—3·20—1·49 and 1·35 per 1000.

Fowl.—One observation. Muscle of breast, 0·50 per 1000.

Grouse.—One observation. Muscle of breast, 0·50 per 1000.

Tortoise.—Two observations. Muscle taken indiscriminately. Figures given—12·11 and 28·62 per 1000.

Frog.—Muscle obtained from two batches of frogs. Figures respectively yielded—5·40 and 9·68 per 1000.

Fish.—Seven observations. Salmon 0·87—turbot 1·98—cod 0·93—eel 0·37—eel 0·36—mackerel 0·33 and plaice 0·34 per 1000.

218   GLYCOGEN AND CLEAVAGE CARBOHYDRATE.

		Amylose carbohydrate per 1000, expressed as glucose.	
Crustacean.—		Muscle of claw.	Muscle of tail.
Lobster	1	4·13	1·62
,,	2	2·58	1·16
,,	3	0·37	0·42
Crab	1	1·77	—
,,	2	6·15	—

PANCREAS.

Dog.—Five observations. Figures respectively yielded—2·21—2·97—0·44—0·75 and 3·20 per 1000.

Cat.—One observation. Figures given—1·47 per 1000.

Horse.—Five observations. Figures yielded—4·44—5·95—5·03—3·62 and 1·20 per 1000.

Sheep.—Five observations. Figures yielded—2·31—2·02—2·74—2·15 and 3·54 per 1000.

SPLEEN.

Dog.—Seven observations. Figures respectively yielded—1·04—2·36—2·14—1·67—0·76—2·07 and 1·18 per 1000.

Cat (spleens from a litter of sucking kittens).—One observation. Figures yielded—2·86 per 1000.

Horse.—Four observations. Figures given—2·52—3·78—2·38 and 0·62 per 1000.

Sheep.—Seven observations. Figures given—4·07—3·27—2·08—2·77—0·94—2·15 and 0·99 per 1000.

KIDNEY.

Dog.—Seven observations. Figures yielded—1·34—1·30—0·51—0·45—1·02—1·13 and 0·57 per 1000.

Cat (kidneys from a litter of sucking kittens).—One observation. Figures yielded—1·97 per 1000.

Horse.—Four observations. Figures given—3·86—1·55—2·13 and 1·54 per 1000.

Sheep.—One observation. Figures given—1·25 per 1000.

BRAIN.

Dog.—Six observations. Figures given—1·00—0·91—0·76—0·36—0·51 and 1·04 per 1000.

Lungs.

Dog.—Two observations. Figures yielded—1·53 and (fœtal pups) 5·02 per 1000.

Horse.—One observation. Figures yielded—4·88 per 1000.

Intestinal Mucous Membrane.

Dog.—Under fasting. Two observations. Figures given—3·70 and 3·10 per 1000.

Under a meat diet. Three observations. Figures given—4·03—3·69 and 3·10 per 1000.

Rabbit.—At a period of digestion. Three observations. Figures given—1·73—2·24 and 2·09 per 1000.

Placenta and Fœtus.

Dog.—One observation upon placenta alone. Figures yielded—1·83 per 1000.

Rabbit.—Four observations. Figures given—placentæ, 22·55—associated fœtuses, 3·12—fœtuses of another animal, 7·37—placentæ and fœtuses together, 2·51 per 1000.

Generative Structures of Oviparous Animals.

Fowl.—Two observations. Ovarian eggs, 4·63, and mucous membrane of oviduct, 11·79 per 1000.

Cod.—Two observations. Ovary (hard roe), 4·73, and testis (soft roe), 2·20 per 1000.

Mackerel.—Two observations. Ovary (hard roe), 3·81, and testis (soft roe), 3·30 per 1000.

Lobster.—Three observations. Ovary, 8·11—ovary, 10·97, and ova adhering to the tail belonging to the same lobster, 8·00 per 1000.

Crab.—Two observations. Ovary with ova, 6·70, and testis with ducts, 3·70 per 1000.

I must repeat what I before stated with regard to the value to be attached to the above figures, namely, that they must only be taken for what they are really worth—that is, as giving a comparative representation of the results obtained from different components of the body by the same process of treatment. Glycogen, when present, may be taken as fully represented. The cleavage carbohydrate from proteid matter is, I know, only represented to the extent of about half the amount that is obtainable by using 10 per cent. instead of 2 per cent. sulphuric acid for conversion into sugar, and grounds exist for believing that there is more carbohydrate material locked up in proteid matter than is brought into view by boiling with 10 per cent. potash in the first instance and the employment of 10 per cent. sulphuric acid for subsequent conversion.

AUTHOR'S CONCLUSIONS.

As the outcome of what has preceded, the views which have prevailed since the promulgation of the doctrine giving to the liver a glycogenic function must be absolutely abandoned. The edifice has been constructed upon a false foundation, and nothing short of its complete demolition will suffice to meet the requirements of physiological truth.

A clearance of the ground is necessary in order that the new material supplied by research may unimpededly be worked into form from a new departure, based upon the sounder methods of experimental procedure placed at our disposal by modern science.

The experimental results I brought forward upwards of thirty years ago, although standing confirmed by others, have not had the full meaning attached to them that I conceive should be given. For my own part, throughout the long series of years mentioned, I have never wavered for a moment with respect to the interpretation the results in question should receive.

Formerly analytical knowledge was not sufficiently advanced to permit of what the experimentalist observed being put into language to secure the production of the same impression upon the mind as would arise from actually seeing the results of the observations themselves. All this is now changed, and quantitative sugar determinations can be expressed with such precision as even to excite surprise at what is attainable. I am constrained, therefore, to think that, in view of the assemblage of facts presented in this volume, the convictions which have so irresistibly taken possession of my own mind cannot fail to penetrate into the minds of others. The error existing has, I know full well, taken deep root in the belief of physiologists, but I have no misgiving that, sooner or later, it will come to be entirely eradicated, and that reference to animal glycogenesis, as it is now understood, will be expunged, except as a matter of history, from works on physiology.

The question is one of great importance, standing as it does at the foundation of the connexion of the carbohydrates with alimentation—in other words, of their disposal within the system—and the comprehension of the nature and origin of the wrong action standing at the foundation of the formidable disease—*diabetes mellitus.* There are two aspects, therefore, presented, giving rise to a demand of a weighty character for exact knowledge. By studying the question from both aspects—physiological and pathological—the investigator is placed in a more advantageous position for compassing its elucidation than when the opportunity is afforded of studying it from one aspect only.

The more extensive the field of observation, the larger the supply of material upon which to generalise. It will not be disputed that the physician and physiologist combined has a wider range of view before him than either the one or the other separately. The physiologist works upon ground where normal action is at play, and seeks to find out how the carbohydrates become naturally disposed of. The physician has to deal with a departure from normal action, taking the form of a failure of power to properly dispose of carbohydrate matter, and to prevent its elimination as sugar by the kidney.

It may, indeed, be truly said that in the consideration of the physiological question I have derived immense advantage from my experience in connexion with diabetes. Closely watching, as I have done for years past, through the quantitative determination of sugar in the urine, the relation between the sugar eliminated and the food ingested, I have been placed in a position I could not otherwise have attained. Without the knowledge thus supplied, I could not possibly have expressed myself in the confident terms in which I consider I am now able to do.

I will now proceed in a concise manner to review the arguments springing from the facts adduced in this work which stand opposed to the validity of the glycogenic doctrine. As carbohydrate matter must, however, of necessity be in some way or other disposed of in the system, we become confronted with the question of the manner in which its disposal actually takes place. Until recently, all I could assert was that carbohydrate matter became assimilated or placed in a position to be susceptible of utilisation within the system, instead of being allowed to reach the general circulation as sugar, and thence to escape as waste material with the urine. By the persevering prosecution of research, however, I am now placed in a different position. The key has been supplied which enables me to follow the carbohydrates along the path taken by them in their application to the purposes of life. Nothing but the disclosures of research could have suggested to me the solution that has been presented. To this matter I will direct attention after the clearance of the ground has been effected from the views that have been hitherto entertained.

Reference to what has been said about the condition of the liver, taken at the moment of death (pp. 133—138), and that of the various other structures of the body (pp. 194—205), shows that it is not true that the liver during life is, as has been alleged, in a more saccharine state than other parts of the system. Practically speaking, the state of the liver under physiological conditions is not essentially different from that of the other structures I have examined; and in the case of muscle it has to be said that the amount of sugar present may be very considerably beyond that found in the liver at the moment of death. Sugar, in reality, exists as a normal constituent of all the tissues and organs of the body. The liver, indeed, stands in the same position as other structures, not only as regards the amount, but likewise as regards the nature, of the sugar present. In the liver taken for examination, without any precautions to prevent *post-mortem* change, glucose is the kind of sugar met with. The sugar, on the other hand, that is found in the liver at the moment of death possesses, as shown by the analyses given (pp. 137—138), a cupric oxide reducing power more or less below that of glucose, in accord with what is seen to be the case in the analyses of the other structures of the body.

Thus one of the main tenets of the glycogenic doctrine is seen to be absolutely untrue, the foundation for it arising from the error of taking the *post-mortem* to represent the *ante-mortem* state.

Reference, further, to the experimental results representing the state of the blood belonging naturally to life (pp. 110—111 and p. 101 *et seq.*) shows that the blood flowing from the liver does not contain more sugar, as it has been asserted to do, than that flowing to it. Sugar is present throughout the contents of the circulation, in which, in fact, it may be considered to exist as a constitutional component, as in other parts of the system. No appreciable difference is found (pp. 166—171) in the amount contained in arterial and venous blood, nor is there less, as has been alleged, in the blood of the portal vein than elsewhere (p. 101 *et seq.* and pp. 158—171). The analyses given show that even after prolonged fasting the portal blood contains sugar similarly to the other parts of the circulation, and that after the ingestion of carbohydrate matter it contains a great deal more. Whilst, speaking precisely, the sugar present in the blood of the general circulation ranges in amount from about 0·6 to about 1 per 1000, the amount in the portal blood after the ingestion of carbohydrate matter may stand as high as 5, and upwards of this, per 1000.

The effect of what has been said is to show that there is no evidence of the occurrence of the disappearance of sugar from the blood in its transit through the systemic capillaries, assumed under the glycogenic doctrine to take place; and, correspondingly, no evidence of the transport, as a functional operation, of sugar from the liver to the systemic capillaries.

The liver, indeed, instead of actually throwing sugar into, or allowing it to pass into, the general circulation, in reality checks the progress of carbohydrate matter onwards. It prevents the fluctuating condition, as regards sugar belonging to the portal blood, from travelling on and being transmitted to the blood of the general circulation.

Through the instrumentality of the liver, then, the blood of the general circulation escapes being influenced by the ingestion of carbohydrate matter. In the exercise of this office of the liver, the sugar from ingestion contained in the portal blood is stopped and converted into glycogen. Thus much is admitted by all, but it is generally thought that the glycogen becomes subsequently reconverted into sugar to be transmitted on, serving only as a convenient form of storage carbohydrate material. This implies that the sugar of ingestion after all reaches the general circulation, and it is contended that the object of conversion into glycogen is merely for the sake of temporary storage—in other words, that at the time of active absorption the carbohydrate matter is detained, to be given out slowly afterwards. Let us see how this argument stands the test of criticism.

Concisely stated, it is asserted that when sugar from ingestion is passing in the portal blood to the liver it becomes, to a greater or less extent, temporarily stopped, to be given back into the circulation when the entry from ingestion is not taking place.

In discussing the point I will, in the first place, direct attention to circumstances existing in the case of the rabbit. In this animal the stomach never becomes empty, as it does in animals generally. Digestion is constantly proceeding, and, even after the lapse of twenty-four hours from the ingestion of food, it is found that the portal blood gives evidence of the presence of absorbed sugar by containing more (*vide* p. 103) than naturally exists in the blood of the general circulation. Assuming, as may fairly be done, that the rabbit does not ordinarily pass longer than twenty-four hours without taking food, it follows that there will always exist in the portal blood more sugar than in that of the general circulation. This is tantamount to saying that under ordinarily prevailing circumstances there is always a state existing calling for the exercise of a stoppage action, which is not reconcilable with stoppage for temporary storage, but, on the contrary, must involve arrest for application in some other way.

From the condition of the urine looked at in relation to different kinds of alimentation, which I will in the next place consider, we obtain a strong point of evidence touching not only the limited question of temporary storage but the whole question of tenability of the glycogenic doctrine.

Healthy urine contains (p. 178) a certain amount of sugar. The amount is in proportion to that existing in the blood of the general circulation (p. 187). The blood and the urine, indeed, stand throughout in accord with each other as regards sugar. When sugar is present to a larger extent than usual in the contents of the general circulation it correspondingly shows itself in the urine. Thus the condition of the urine in relation to sugar moves with, and in fact serves as an index of, that of the blood. Through the urine, therefore, knowledge is supplied concerning the quantity of sugar that is permitted to reach the general circulation.

Now, with the animal feeder, in which the ingestion of only a limited amount of carbohydrate occurs (and even it may be said, in the fasting animal where no ingestion is occurring) the urine stands in the same position with respect to sugar as that of the vegetable feeder, whose food contains a large amount of carbohydrate matter. The blood of the general circulation, in conformity with what has previously been said, likewise corresponds in the two cases. With the ingestion of the limited amount of carbohydrate, the blood and the urine contain a certain amount of sugar. With the ingestion of an amount incomparably greater, the same state of things is still encountered. Both the blood of the general circulation and the urine remain uninfluenced by the ingestion of carbohydrate matter, so long as it does not exceed ordinary limits. But stoppage from entry into the circulation, through storage by the liver, cannot be indefinitely carried on, and, taking into account a lengthened period of time, it is obvious that, if ingested carbohydrate were in reality transported as sugar to the systemic capillaries, its transit in the case where there has been free ingestion must proportionately add to that existing where the ingestion has only been limited in amount. Sooner or later the extra amount must pass through the circulation, and where from day to day the quantity to pass is ten, twenty, or thirty fold greater than under an animal diet, the addition must present a source for the kidney to draw upon and thereby lead to a proportionate presence of sugar in the urine.

With the limited ingestion associated with animal food, the urine contains, as has been said, a certain amount of sugar. Any additional carbohydrate ingested should have the effect of augmenting this sugar, and of doing so in proportion to its extent. It is not, however, in reality found that under ordinary circumstances the urine gives evidence of being influenced by carbohydrate food. Hence, for compatibility, we have to say that the passage of any quantity of sugar into the circulation from carbohydrate matter ingested within ordinary limits goes for nothing as regards effect upon the urine, whilst influence is exerted by the known small amount of sugar existing in the blood in association with an animal diet and even with an absence of food.

The liver, in fact, instead of doing what is claimed for it under the glycogenic doctrine, does exactly the reverse. It neither forms sugar to be discharged into the general circulation and conveyed to the tissues, nor temporarily stores up carbohydrate matter from ingestion to be subsequently allowed as such to pass on. On the contrary, it keeps the general circulation free from the sugar that would otherwise enter and show itself in the urine. Its office is an arresting one in relation to carbohydrate matter, and if it were not for the exercise of this office we should all be in the same position as the diabetic. It furnishes a line of defence against the passage of carbohydrate in a free state into the circulation, and thus prevents the sugar, derived from alimentary absorption, which is contained in the portal blood, from proceeding further. In proportion as the line of defence is ineffectual, sugar will reach the general circulation and then the urine.

As an outcome, the difference between health and diabetes may be expressed in the following terms :—

In health the capacity exists of stopping the onward progress into the general circulation of the sugar derived from ingested carbohydrate matter, when the ingestion stands within ordinary limits. When the ingestion exceeds ordinary limits the capacity of arrest, which is not unbounded, may not be equal to effecting a complete stoppage. A portion may in this way reach the general circulation, and, in proportion as it does so, will appear in the urine. Hence the explanation of the saccharine urine found to occur (p. 116) as an accompaniment of excessive feeding with sugar. In further illustration, I may mention that the highest figures I have obtained for the sugar in normal blood have been with rabbits which had been highly fed with oats. In accord, the urine in these instances has given evidence through the ammoniated cupric test of containing a larger amount of sugar than is ordinarily met with in healthy urine.

In diabetes, on the other hand, carbohydrate matter is not properly stopped and disposed of, but is permitted, instead, to reach the general circulation in the form of sugar. The faulty state presents itself in all stages of advance. Every grade of diversity exists between the healthy state and the state belonging to the severest form of diabetes. In some instances the capacity of stoppage is only just below the normal. In these it is only when carbohydrate matter is rather freely ingested that saccharine urine is encountered. In others more and more impairment of the power to check transmission into the general circulation exists, and in correspondence less and less carbohydrate matter can be ingested without appearing in the urine.

One point is linked with another, and my experience of diabetes counts largely as a help towards enabling me to speak with the precision and confidence I am doing. Besides a large number of cases seen in the course of my many years' hospital practice and in private consultations outside my house, I have, extending up to the moment at which I am writing, a record in my case-books of 2,642 cases seen at home, in all of which I have closely watched, under the most varied circumstances and through the quantitative examination of the urine, the elimination of sugar in relation to the food ingested.

Now, the teachings of this experience point unequivocally to its being contrary to the physiological order of events for ingested carbohydrate to reach the general circulation in a free state to be transmitted to the systemic capillaries for destruction in some, it must be said, unknown way. No other conclusion, I hold, is permissible, and I have no misgiving that sooner or later this view will meet with general acceptance. In fact, it is through the occurrence of what the glycogenic doctrine implies that diabetes is caused, and the symptoms of the disease are due to sugar being allowed to reach the general circulation. In proportion as it does so, a deviation from the natural state is created by the sugar of the blood being raised from its normal standard proportion.

The effect of dietetic treatment in diabetes stands in strict harmony with the view adopted. As I have already said, the disease essentially consists of a loss, or of more or less impairment, of the power which naturally disposes of ingested carbohydrate matter, and prevents its reaching the general circulation in the form of free sugar. Of the nature and effect of this power I shall speak later on.

Through the defective power existing in diabetes, sugar finds its way into the blood of the general system, and, in proportion as it does so, places it in an unnatural state, the effect of which is to interfere with the performance of nutritive action, and of the processes of life generally in a healthy manner. From the position held by the blood, it is only in accord with what may reasonably be expected that with an altered constitution it should, in proportion to the extent of alteration, carry disturbance everywhere.

The deviation from the natural state, induced by ingested carbohydrate matter being permitted to reach the general circulation, to which the various troubles belonging to diabetes, it must be considered, are attributable, depends as regards degree, in the first place, upon the extent to which the impairment of the power of stoppage exists, and, in the next, upon the amount of carbohydrate ingested. The one is at the root of the disease, and if it could be rectified all would immediately be set right. The other is under our control, and by proper regulation of the diet much may be done towards ameliorating the condition existing. It must be remembered that animal as well as vegetable food contains a certain amount of free carbohydrate; and, as I have shown in this work, sugar is set free by digestion as a cleavage product from proteid matter. Therefore, in all food a supply of carbohydrate matter is given from without. This, in part, accounts for the sugar encountered in severe cases where sugar is passed, notwithstanding restriction to a purely animal diet.

I have said "in part" accounts for the sugar encountered; for, in severe cases, there must be something further to be dealt with, seeing that sugar is even passed apart from the influence of food. There can be no doubt that sugar is susceptible of being derived from the tissues, and, looking at what I have been led to propound with regard to the glucoside constitution of proteid matter, there is nothing unintelligible about it. On the contrary, in view of all the circumstances, it is a reasonable supposition that in connexion with the altered state of things existing, a ferment becomes present endowed with the power of wrongly splitting up the glucoside proteids of the body in the same manner as, for instance, amygdalin is split up by emulsin.

There is, then, a class of case in which the fault consists only of a loss, or, it may be, varying degrees of impairment, of the power of disposing of ingested carbohydrate matter in such a manner as to prevent its reaching the general circulation; and another, in which, in addition to this, a condition within exists attended with the splitting up, with sugar as a cleavage product, of the proteids of the body. The former is completely controllable by dietetic management, the latter only so to a partial extent.

Most cases, I will not go so far as to say all, are in the incipient state controllable, with respect to the elimination of sugar, by diet. In young subjects it is usual to find that the condition is observed to remain only temporarily controllable. An insidious "something" advances, carrying the case on into a more and more pronounced state of development. In subjects of more mature years, on the other hand, if the morbid tendency is not encouraged by allowing the unnatural condition, arising from the continued pervasion of the system with sugar derived from defectively disposed of ingested carbohydrate, to exist, it is usually observable that the case may be kept indefinitely controllable. I may, even, go further and say that, under strict and persevering attention to dietetic management, conjoined with a certain form of drug treatment which I believe affords material help, experience shows that a tendency to restoration is promoted; and that, as a result, whilst at first no starchy or saccharine food could be taken without giving rise to the elimination of sugar, in the course of time more and more power of properly disposing of carbohydrate matter is acquired, leading to the toleration of more and more carbohydrate matter in the food without giving rise to saccharine urine until, it may be, the capacity for the resumption of an ordinary diet is re-established.

It may be safely accepted that the point to be attained for promoting the restoration of the power of properly disposing of carbohydrate matter is to establish and maintain as close an approximation as possible to the natural state as regards the blood, and thereby the system throughout, in relation to the presence of sugar. On bringing the position to a normal one in this respect, all the symptoms of the disease at once disappear, whilst previously it may have been noticed that they were more or less speedily growing.

I have seen it asserted, under the erroneous notion of there being a functional transport of sugar to the systemic capillaries, that carbohydrate articles of food ought not to be totally withheld even although a considerable voidance of sugar may be occurring. This is equivalent to saying that a deviation from the normal state as regards the blood and the system throughout is to be disregarded. Moreover, even if the premises were sound, the argument is illogical, as the voidance of sugar affords proof that already the blood is surcharged with sugar which is running off as waste material. To add more through the medium of the diet would be simply to increase the surplus existing. It is true, through failure of appetite and inability to take sufficient food under restriction to the list of appropriate articles, reasonable grounds for relaxing the diet may arise, but this is a totally different matter.

Restriction from carbohydrate food should not be carried further than experience shows to be necessary, in the particular case under consideration, to maintain a normal state of urine. The amount of carbohydrate matter that can be taken without leading to the voidance of sugar presents considerable variation in different cases, and it happens that if what can be utilised is not supplied, a loss of bodyweight and failure of strength ensue. Often the complaint sets in with only a slight diminution of the natural assimilative power, presenting itself under the form of what is spoken of as glycosuria. Under such circumstances only a moderate restriction of diet is needed to meet what is wanted. If more be enforced, loss of weight will probably follow and the appearance be given that the treatment is inflicting harm. By extending the liberty to that which the state of the urine shows may be done without over-stepping the range of toleration, immediate improvement will occur. Such, in every case, should constitute the principle of action to be adopted.

It may happen that restriction in diet is needed to be enforced to the full extent at the commencement of treatment, and not so later on. An interesting sequence is noticeable in these cases. The patient, it may be, has been the subject of diabetes for some time without knowing from what he was suffering. The nature of his malady becomes discovered, and he is placed, as a part of the treatment adopted, on the restricted diet. He had been hitherto steadily losing ground, but now his symptoms disappear; he gains in weight, and, altogether, in the course of a short time he becomes restored to a greatly-improved condition. The treatment has at once produced a marked effect upon the urine, but perhaps a little time has been required to get it entirely free from sugar. It remains now constantly free, and the patient continues to improve in health and strength. After a while it may happen to be noticed that a change begins to set in. Notwithstanding that the urine continues free from sugar, the treatment, which before produced improvement, is now attended with a retrogression in weight, and naturally the patient's mind becomes disturbed from apprehension of a return of his disease.

Instead, however, of ground for concern existing, my experience informs me that the loss of weight under the circumstances mentioned is to be taken as an indication of restoration, to a greater or less extent, of the power of properly disposing of carbohydrate matter. The power existing, and the opportunity not being given for its exercise, the want appears to be felt by the system. At all events, in these cases I have confidence in concluding that on trying a little starchy food (best, I consider, in the form of ordinary bread, as the loss of this may be regarded as constituting the greatest privation) it will be found not to give rise to the appearance of sugar in the urine. On this proving to be the case, the amount should be guardedly and gradually increased, short of leading to a return of sugar. The result noticeable will be a speedy recovery and subsequent maintenance of weight.

Remarkable, but nevertheless correct, to state, in cases where a loss of weight does not, under corresponding circumstances, supervene, my experience justifies the inference that the power of utilising carbohydrate matter has not become raised, and that, consequently, the ingestion of starchy food will be followed by the appearance of sugar in the urine. Thus, when the power of making use of starchy food has been restored, and starchy food is not given, a loss of body-weight is observed, whilst such is not noticeable where a restoration of power has not occurred. Information is here supplied which in practice I turn to account as a guide in the details of dietetic management. Every case requires to be dealt with upon its own merits, and for successful treatment a rational and intelligent plan of procedure, founded upon a precise knowledge of the condition of the urine, must be adopted.

The only point remaining for consideration is the manner in which carbohydrate matter becomes disposed of in the system. This really is the most important point of all in connection with our subject, as it deals with the purposes to which the carbohydrates are applied in the economy of life.

The glycogenic doctrine throws no light upon this matter. It falls short of reaching the point. It assumes that carbohydrate matter is thrown into the general circulation in the form of sugar for functional transport to the systemic capillaries, but it fails to proceed further, and to show in what manner the sugar is disposed of. It simply leads into darkness, and nothing but fruitless efforts have been made to obtain light in the direction looked for. This is not, it may be said, to be wondered at, seeing that there is not the functional transport that has been assumed to occur; and that the facts before us show that we must seek for the disposal of carbohydrate matter before the opportunity is afforded of its reaching the general circulation. To attempt to discover the manner of disposal in a situation that the sugar does not naturally reach is not likely to be attended with any satisfactory issue, and those who continue to prosecute enquiry upon such a foundation can scarcely fail to be left endlessly groping in the dark.

There is no living matter without proteid, and probably even the broader statement may be made, that there is no living matter without proteid, carbohydrate, fat, and certain mineral constituents. Carbohydrate matter may thus presumably be looked upon as universally diffused through the living kingdom of nature; and, so circumstanced, it must have a comprehensive part to play in the economy of life. In speaking of the manner in which it becomes disposed of in the animal system, it seems to me that this consideration should be held prominently in view.

If we look around and give attention to what is happening, it is observable that certain changes are wrought upon carbohydrate matter by the agency of living protoplasm. By virtue of the power with which living protoplasm is endowed, carbohydrate matter located within its sphere of influence is noticed, it may be said, to undergo:—

1. Transmutation;
2. Application to the production of proteid; and
3. Transformation into fat.

The capacity for producing these effects may be regarded as constituting a common property appertaining to protoplasm in an active state, quite irrespective of which of our conventional divisions of living nature the protoplasm belongs to. Universality of action is traceable without distinction in the animal and vegetable kingdoms.

I will proceed to speak in detail of the disposal of carbohydrate matter by protoplasmic action under the respective heads enumerated above. From what has to be said upon the subject, I think it will be seen that we need not seek further for the explanation that is wanted to place the matter in a clear and intelligible light. Not only do we find that an adequate explanation is presented to account for the disposal of carbohydrate within the system, but that the seat of disposal is located just where it properly should be, in order to permit of escape from the production of a diabetic state—that is, before the entrance to the general circulation is reached.

1.—*Transmutation.*

The transmutation of carbohydrate matter was dealt with at some length in an early part of this work (p. 18, *et seq.*), and it was seen that, whilst ferments and chemical agents transmute by increase of hydration—carrying, for instance, the principles of the amylose into the saccharose and glucose groups—the effect of the influence exerted by living matter is to reduce from the higher to the lower states of hydration. For example, the sugar contained in the sap of the plant is observed to be transmuted into cellulose, which constitutes a fabric material of the organism, and into starch, inulin, &c., which serve as storage materials, and, in reality, contribute in the end to fabric construction.

Through transmutation, carbohydrate matter meets with its chief application in the vegetable kingdom, the cellulose from which the fabric of the plant is constructed constituting an ultimate product. I do not know whether it is to the full extent realised that the storage carbohydrate of the seed, tuber, &c., passes in reality to the same destination and that the process of storage only represents a temporary halt on the way. By protoplasmic action certainly, and probably by incorporation with nitrogenous matter into proteid and transit through protoplasm, the sugar of the sap becomes transmuted, apparently in each case in a similar manner, into fabric and storage material. The storage carbohydrate, when exposed to conditions to lead on to its application, becomes first of all hydrolysed by ferment action. Thus reconverted into sugar, it again stands in the position it held in the sap from which it was taken for storage deposition; and now by the exercise of protoplasmic action in the developing organism it is carried as a final step into cellulose, for it is from the starch of the seed, &c., that the cellulose of the growing structure is derived.

In the animal kingdom evidence is adducible of the occurrence of transmutation in an analogous manner to what is observable in the vegetable kingdom. Transmutation, however, does not here to a like extent lead on to application to texture construction; indeed, it is only to quite an insignificant extent that it does so. An example is forthcoming in the deposition of cellulose as a texture basis of the test or outer investment of the tunicata.

Transmutation, however, comes somewhat largely into operation in reducing sugar to the state of a carbohydrate of lower hydration for storage purposes. Glycogen is the form of carbohydrate into which the sugar is in the animal transmuted, and it is interesting to notice that in the tribe of organisms of the vegetable kingdom, which are wanting in some of the attributes of plants, and present exceptional analogies to the animal, namely, the fungi, the storage carbohydrate exists under the form of glycogen and not of starch.

A notable illustration in the animal kingdom of transmutation by reduced hydration for storage is supplied by what occurs within the liver. The sugar contained in the portal system, taking origin from the carbohydrate matter ingested, is stopped by the cells of the organ and transformed into glycogen. By transmutation into glycogen the carbohydrate is checked in its onward progress to the general circulation, and subsequently, it may be reasonably assumed from all the knowledge at our disposal, meets with application through one or both of the other methods of disposal of carbohydrate matter by the agency of living protoplasm.

It is interesting to note that in tracing the sequence of events occurring in the animal system in connection with ingested starch, we have the same train of phenomena to deal with as in the growing seed. The starch is first transmuted by ferment action—by diastase in the one case and the amylolytic ferments of the digestive secretions in the other—into sugar, a body susceptible of diffusion and of transport in solution. By the portal system the sugar is conveyed to the liver and brought within the sphere of influence of its cell-protoplasm. Here transmutation of the opposite kind to that before occurring takes place, with the production of glycogen. By diffusion the sugar taking origin from the starch in the growing seed arrives within the reach of the power belonging to the living protoplasm of the embryo, and, as in the case of the liver, transmutation in the direction of diminished hydration occurs, resulting here in the production of cellulose. The same train of phenomena may be traced as constituting a part of the life-history of the plant itself. The primordial starch produced by the chlorophyll corpuscles in the leaf reaches the sap as sugar, which on arriving at seats of protoplasmic activity is transmuted down in hydration to fabric and storage carbohydrate. The sap of the plant, looked at from the point of view indicated, stands, it will be observed, in an analogous position to the contents of the portal system of the animal.

I have referred to the protoplasmic transmutation of sugar into glycogen in the liver. But the transmutative power by no means exclusively belongs to the liver—indeed, I am disposed to think that it exists as a general property of the protoplasmic matter of the body. Certainly, glycogen is recognisable, and even it may be largely so, in other parts of the body, and it is probably produced at its seat of presence. Every part of the body contains sugar intrinsically belonging to it, which may come from the cleavage of the proteid matter around. The glycogen may be derived from this sugar, or may possibly constitute a product of the cleavage process. In the multifarious actions occurring in a living part there is, doubtless, much complexity of result arising from the antagonistic effects of ferment and protoplasmic actions.

Further evidence is afforded through the kind of sugar found in the urine after the direct introduction of glucose into the system of the carbohydrate-transmuting power of living matter in the direction of a diminution of hydration, taken as represented by a diminished cupric oxide reducing capacity.

At p. 189 I gave a description of experiments in which glucose, derived from honey and proved by examination to have the same cupric oxide reducing power before and after sulphuric acid, was injected into the jugular vein. In two out of the three instances in which urine was procurable the sugar eliminated was found to be in a state considerably removed from glucose, the initial cupric oxide reducing power belonging to it being considerably less than that shown to exist after treatment with sulphuric acid. The results of other experiments were also given, conducted with lævulose derived from the recently introduced trade article. Here, however, the destructive action exerted upon the sugar by boiling with sulphuric acid, in accordance with the known effect upon lævulose, was such as to lead to considerably lower figures being obtained after than before treatment with the acid. From this counteracting circumstance no opportunity was given of ascertaining whether in reality any diminution of cupric oxide reducing power occurred.

It would hardly have been surmised that such results would have been yielded. They agree, however, with the general tenor of experience set forth in this work which is to the effect that by the agency of living matter the carbohydrates are moved within the system in the direction of lessened cupric oxide reducing power instead of in the reverse direction as happens as a consequence of ferment action. Thus fitting in with and supporting, as they do, a general principle of action shown by other evidence to be in operation, they acquire considerable importance. On this account, in order that the point might be placed beyond doubt, I deemed it advisable to undertake a further series of experiments, and I found that I could simplify the operative procedure by injecting the sugar solution into the subcutaneous tissue instead of into a vein. As will be seen from the results to follow, confirmatory evidence was supplied, and it may be considered to be established that sugar reaching the general circulation as glucose, and thence brought into contact with living protoplasmic matter whilst contained in the vessels and whilst passing through the secreting structure of the kidney, escapes in a form possessed of a lower cupric oxide reducing power than glucose.

The list comprises seven experiments conducted upon rabbits consecutively taken. In each case 1 gram of glucose (derived from honey) per kilo. of body weight was injected into the subcutaneous tissue of the back, the animal being temporarily placed under ether for the performance of the injection. Urine was obtained from the bladder about two hours afterwards. The delicacy of the ammoniated cupric test is such as to permit of very small quantities of urine sufficing for analysis, and dilution to the extent of 30, 40, or 50 times is requisite. The urine, suitably diluted, was divided into two portions, one being titrated at once, and the other after boiling for an hour and a half with 2 per cent. sulphuric acid under the inverted condenser. In each case, as shown by the tabular arrangement given of the results, the sugar present in the urine possessed a cupric oxide reducing power much below that of glucose, although glucose was the form of sugar injected.

Nature and amount of Sugar in the Urine after the subcutaneous injection into the Rabbit of 1 gram of glucose per kilo. of body weight.

		Sugar per 1,000, expressed as glucose.	Cupric oxide reducing power of the sugar eliminated in relation to that of glucose at 100.
Rabbit I. 10 c.c. of urine obtained.	before sulphuric acid after ,, ,,	48·08 65·79	73
Rabbit II. 8·6 of urine obtained....	before sulphuric acid after ,, ,,	19·84 51·89	38
Rabbit III. 3 c.c. of urine obtained..	before sulphuric acid after ,, ,,	39·68 94·46	42
Rabbit IV. 4 c.c. of urine obtained..	before sulphuric acid after ,, ,,	34·72 69·44	50
Rabbit V. Over 15 c.c. of urine obtained............	before sulphuric acid after ,, ,,	21·74 35·72	61
Rabbit VI. Over 30 c.c. of urine obtained............	before sulphuric acid after ,, ,,	10·41 19·23	54
Rabbit VII. 4·1 c.c. of urine obtained.	before sulphuric acid after ,, ,,	48·78 110·85	44

The above results very strikingly show that the amount of cupric oxide reducing power after treatment with sulphuric acid greatly exceeded that existing before. It is asserted that animal gum, an amylose carbohydrate, may exist in the urine. If present, it would serve to account for the difference revealed in the results obtained. To ascertain if the difference were attributable to such a source, a portion of the urine in three of the experiments was poured into a large quantity of absolute alcohol. The effect of this upon animal gum would be to precipitate it. After standing for twenty-four hours the alcohol was filtered off, evaporated down, and the product subjected to titration with the ammoniated cupric test before and after treatment with sulphuric acid. The results obtained revealed the same kind of difference as was noticed in the original examination, thus rendering it evident that the phenomenon observed must be ascribed to the presence of sugar with a lower cupric oxide reducing power than that of glucose.

I have mentioned that a certain amount of sugar exists in normal urine, and the evidence before me points to the nature of the sugar standing in accord with what has been mentioned above in exerting an increased cupric oxide reducing action after treatment with sulphuric acid.

The kind of sugar eliminated in diabetes is, on the other hand, ordinarily glucose. I have had a number of examinations made, and it is only in a few instances that a sugar with a lower cupric oxide reducing power has been met with.

2. *Application to the Production of Proteid.*

The observations of Pasteur upon the growth of yeast may be considered to reduce to demonstration the production of proteid by the agency of protoplasmic action from the incorporation of carbohydrate with nitrogenous matter. The question was considered in an early part of this work (p. 53, *et seq.*), and it was there shown that yeast cells, which consist of little masses of living protoplasm, placed in a medium composed only of sugar, tartrate of ammonia (nitrate of ammonia, which has no carbon entering into its constitution, and, therefore, renders it absolutely evident that the carbon of the newly formed proteid must be derived from the sugar, may be substituted for the tartrate), mineral matter, and water, grow and multiply—a fact which implies that fresh protoplasm, and hence proteid, must be produced. The conditions here are so simple as to leave no room for hesitation in accepting the conclusion to which they lead up.

Upon the strength of what is observable in the higher vegetable organisms, it is likewise affirmed in settled terms by vegetable physiologists that carbohydrate matter is utilised in the production of proteid by incorporation with a nitrogenous principle (*vide* pp. 54 —56). Asparagin, a crystallisable and diffusible nitrogenous principle widely dispersed through the vegetable kingdom, appears to play an important part in connexion with the operation. Sachs, in his work on the 'Physiology of Plants,' in several places speaks of it as an established point that from carbohydrate matter and asparagin proteid is formed, and further asserts that the converse process of splitting up of proteid with the liberation of asparagin, for fresh service in the same direction, likewise constitutes an operation occurring in association with metabolic activity. Other nitrogenous principles may take the place of asparagin, and with the supply of the nitrogen-containing compounds, on the one hand, from the soil, and the production of carbohydrate in the leaves from the carbonic acid of the atmosphere and the elements of water through the influence exerted by the sun's rays, we have before us the source of the principles entering into the constitution of proteid matter.

It is, then, distinctly affirmed that proteid in the plant is constructed from carbohydrate matter and asparagin, or some other nitrogenous principle, by the agency of protoplasmic power, and, further, that a dissolution of the proteid thus formed afterwards takes place, with the liberation of asparagin, which again enters into the construction of proteid by conjugation with fresh carbohydrate matter. Nothing is here said by Sachs about carbohydrate issuing as a cleavage product from the proteid, although it is spoken of as constituting the complementary part to the asparagin in the process of construction. Elsewhere, it is true, reference is made to protoplasm being concerned in the deposition of starch, &c., as storage, and cellulose as fabric material, but the question is not pursued to any further extent. Looking, however, at all the knowledge in our possession, I am led to go on and say that the weight of evidence and probability is in the direction of proteid construction constituting an important intermediary in the physiological progression of carbohydrate matter. Through its glucoside constitution, proteid matter stands as an important factor in the play of changes connected with life, into which the carbohydrates enter.

Let me carry the discussion into the domain of materiality, and take for illustration a growing bud. At the seat of metabolic activity asparagin is produced by the dissolution of previously formed proteid matter. The asparagin thus originating exists in contact with sugar continuously being derived from the starch taking its source in the chlorophyll corpuscles under the operation of the solar energy. Through the instrumentality of the living protoplasm within the sphere of influence of which the two principles are lying, incorporation into proteid takes place with the contemporaneous growth of protoplasm. Observation testifies that where the protoplasm has existed, fabric cellulose becomes deposited, and, from a review of all the circumstances, it may be looked upon as probable that the deposition occurs as the result of proteid cleavage, with the concurrent liberation of asparagin to contribute to a repetition of the operation with fresh carbohydrate matter, generated by the starch-forming chlorophyll. Under the view presented, the protoplasm of the growing part incorporates into itself the newly formed carbohydrate, and subsequently leaves it behind as lignified cellulose in its track, just as a coral polype incorporates into its substance calcareous matter which it leaves behind in the form of the coral structure as it grows on.

The account that has been given will equally apply to the growth of the seed and the shoot of the tuber, the only difference being that the new organism draws its carbohydrate from storage material instead of from a generating source.

Economy of nitrogenous matter is noticeable in the process described. A certain amount, it is true, of the protoplasmic nitrogen is left behind with the fabric carbohydrate, and in connection with the storage carbohydrate there is evidently a special deposition of nitrogenous matter to meet the requirements of the growth subsequently to take place; but with the repeated service of asparagin that has been alluded to, an economical provision, as regards nitrogenous matter, exists for bringing about the utilisation of a large amount of carbohydrate material.

I have drawn proof of the production of proteid by the incorporation of carbohydrate with nitrogenous matter through protoplasmic power, from the actions occurring in the vegetable kingdom. According to the view I am propounding, the disposal of carbohydrate matter in the several ways to which I am referring is the result of an attribute or power belonging to living protoplasm, without distinction as to which of the conventional kingdoms of nature it happens to fall into. I will proceed to adduce evidence pointing to the utilisation of carbohydrate in the animal kingdom in the production of proteid matter; but, before doing so, I may remark that what has been said in this work about the glucoside constitution of proteid matter gives very great support to the whole proposition. Utilised in the formation of proteid, it is only in accord with what might be expected that it should be susceptible of withdrawal under the operation of influences effecting a disruption of the proteid molecule. The two stand in harmony, and mutually substantiate each other.

We start with the carbohydrate matter derived from ingestion in the alimentary canal. The chief portion takes rise from free carbohydrate contained in the food, but a portion (p. 50) issues from the cleavage of the proteids of the food by the proteolytic ferments of the digestive tract. From whatever source starting, carbohydrate matter in the form of sugar is lying in association with the nitrogenous product of proteolytic ferment action—peptone—within the sphere of influence of the extremely active little masses of protoplasm constituting the cells covering the villi. The circumstances here present no essential difference from those existing in Pasteur's observation. For the yeast cells we have only to substitute the cells of the villi; and, one kind of protoplasm having the power of conjugating carbohydrate and nitrogenous matter into proteid, it may not unreasonably be looked for that the other should also be endowed with it.

I have shown that the portal blood, after the ingestion of carbohydrate food, contains considerably more sugar than is met with in other parts of the circulatory system. From the application of the analytical procedure for obtaining cleavage carbohydrate from proteid matter, it is learnt that higher figures are here also given than by the blood existing elsewhere. This matter is referred to at p. 215 et seq., and it is there further seen that, after the injection of sugar into the jugular vein of rabbits, the contents of the circulatory system furnished considerably larger amounts of amylose carbohydrate than under other circumstances has been found. The evidence upon the point, taken in its entirety, certainly tends to show that the presence of carbohydrate leads to increased results being yielded by the analytical procedure for obtaining cleavage carbohydrate, thus falling in with the view propounded regarding the application of carbohydrate to proteid formation.

What is noticed with regard to peptone stands in harmony with the disposal under consideration of carbohydrate matter. By the proteolytic ferments of the digestive system, the proteid molecule is split up with the production, as is well known, of peptone; and, as I have shown in an earlier part of this work (p. 50), of sugar. This is the first step in the application of proteid matter as food. Thus broken up into more simple principles, it is placed in a suitable state for the exercise of the synthetic action of the living protoplasm within the sphere of influence of which the products lie. Peptone may be regarded as standing in the position of asparagin and other allied nitrogenous principles of the vegetable organism, and also, it may be said, of the tartrate of ammonia in Pasteur's yeast culture liquid, in relation to the production of proteid. Its incorporation with carbohydrate matter by the agency of the synthetic power with which living protoplasm is endowed, suffices to account for the production in the same manner as occurs in living nature generally of the proteid matter into which, by common consent, it is conceded to pass. Nowhere, perhaps, does more active protoplasm exist than in the investing cells of the villi; and it is doubtful whether sufficient attention has been given to their importance as protoplasmic instruments of action in the economy of the living animal. The altered characters they present during digestion, as compared with fasting, testify to their activity. Their number must be extremely great, and if, instead of being spread over a free expanse of surface, they were packed into a glandular mass constructed after the usual fashion, a good sized organ would result. The position they occupy does not in reality detract from the exercise of an assimilative office as effectually as if they were arranged in a compact gland; and any other position than that of being spread around the alimentary tract would be incompatible with applicability to meet the requirements associated with food.

Now, peptone disappears from view just where it should do under the view that it is utilised in the production of proteid by the agency of the cells of the villi, or, speaking more generally, the cells belonging to the inner surface of the intestinal canal. The presence of peptone is easily demonstrable within the alimentary canal, but it is not to be discovered in the contents of the general circulation, nor even in the portal blood or chyle. Its disappearance is spoken of by physiologists as being involved in a certain amount of mystery; and it is stated that all that can be definitely said is that the mucous membrane constitutes the seat of its conversion into proteid. Giving, however, to the investing cells of the villi the same power that protoplasm existing elsewhere is endowed with, there is nothing mysterious in the disappearance of peptone where observation shows that it occurs. Indeed, with the presence of peptone and sugar in the alimentary canal, and the synthetic influence of protoplasm within reach to exert its action upon them, the disappearance of the former by incorporation into proteid is only what might be looked for. The chief channel through which the newly-formed proteid is conveyed into the system is probably the lacteals; but, apparently, a certain proportion reaches the portal blood.

Thus, the points that have been referred to respecting peptone give support to the view that has been propounded with regard to the application of carbohydrate matter to proteid formation through the medium of the protoplasmic power of the cells of the villi.

Other considerations of a more general nature tend to show that the nitrogenous portion of our food is concerned in the assimilation of carbohydrate matter. Where it is deficient, evidence is afforded that the carbohydrates do not become applied within the system in the same beneficial manner as where it is present in due proportion. The food must contain a certain quantity of nitrogenous matter for the carbohydrate portion of it to be turned to proper account. The absorbed material which escapes application by the protoplasm of the investing cells of the villi reaches the portal blood and thence the liver; and it is found that in proportion as carbohydrate food is in excess, so does the liver become charged with transmuted sugar under the form of glycogen. In the laboratory, advantage is taken of this fact when it is desired to procure glycogen in quantity.

I look upon it that there can actually be no doubt that one of the purposes to which the carbohydrates are applied in the economy of life is participation in the production of proteid matter. Such being the case, the position assigned in Liebig's classification to the nitrogenous constituents of food as exclusively representing the flesh-forming principles, can no longer be considered to hold good. With incorporation into the proteid molecule, carbohydrate matter contributes in reality to flesh formation; and, universally distributed throughout the system in this state, it represents a large amount of locked-up carbohydrate existing in the body. Thus circumstanced, it stands in a position to permit of its retention, and possible further utilisation, in contrast to the condition existing in diabetes, where, reaching the general circulation in a free state, it passes off as unused material with the urine.

A few pages back I spoke of proteid, through its synthesis and cleavage, constituting a medium for the utilisation of carbohydrate matter in the vegetable kingdom. The same probably occurs in the animal kingdom. For instance, on looking at the various surroundings of the case, the idea, it seems to me, that most satisfactorily accounts for the origin of the lactose, fat, and casein of milk is that they all constitute cleavage products from proteid matter. Casein, it happens, as I mentioned (p. 32) when speaking of the glucoside constitution of proteid matter, differs strikingly from the other proteids examined in yielding quite an insignificant amount only of cleavage carbohydrate— a condition that harmonises with what might be looked for on the hypothesis of its cleavage origin. The sugar met with in the several structures of the body may take origin from proteid cleavage. In the advanced stage of severe cases of diabetes, some of the sugar eliminated is evidently drawn from the tissues of the body. An excessive cleavage of proteid, arising from the particular state existing, adequately affords the explanation wanted.

(3). *Transformation into Fat.*

The question of whether it is possible for carbohydrate matter to be transformed into fat in the animal system was the subject of warm controversy during the first half of the present century, and several names of renown stand associated with it. The settlement then arrived at was in the affirmative, and the advance of knowledge that has since taken place attests the correctness of the conclusion. The animal kingdom, indeed, it may now without hesitation be asserted, stands in the position of a large constructor of fat from the carbohydrate matter constituting the primordial organic production supplied to the living kingdom of nature by the vegetable organism operating under the influence of the power derived from the sun.

If it be true, as is believed, that carbohydrate matter represents the initial condition of all organic products, the whole of the fat encountered in both kingdoms of living nature must take its origin directly or indirectly from carbohydrate matter. From the nitrogen-containing compounds and mineral matter supplied to the plant through the roots, and the carbohydrate matter produced by the leaves, we have the basis for the construction of the various complex organic products that exist. Now, evidence is adducible from which the conclusion may be drawn that fat is susceptible of taking origin from these complex organic products by dissociation, brought about in the presence of certain conditions. We do not in this case discern, upon a superficial view of the matter, any direct connexion between fat and carbohydrate, but, if carbohydrate is the source of fat and all the carbon compounds of living nature, the fat must, whatever the intermediate stages that may have existed, have ultimately taken origin from it.

From what has been said, however, in this work about the origin and the constitution of proteid matter, the sequence of events becomes intelligible. In view of chemical experience, there is nothing unreasonable in the proposition that carbon may enter the proteid molecule as carbohydrate, and come out from it in the form of fat. Under one set of conditions the cleavage may be attended with the liberation of a carbohydrate, and, under another, with that of fat. Even with the direct production of fat from carbohydrate, there may be passage through proteid matter. Certainly, protoplasm is a necessary factor in the process, and there are grounds for considering it more than probable that the result is attained through incorporation and liberation, rather than by direct transformation. Should the former supposition be right, there is no essential difference between the production of fat, as an act of assimilation, from carbohydrate and the origin of fat, by the cleavage of proteid matter belonging to the textural constituents of the body, which takes place to a prominently marked extent in fatty degeneration.

The growth of yeast cells in Pasteur's sugar pabulum has been already taken advantage of to illustrate the transmutation of carbohydrate matter from a higher to a lower state of hydration, and its application, in the presence of a nitrogen-containing compound, to the construction of proteid. The example in question may also be made use of in connexion with the further point that is under consideration, inasmuch as it likewise serves to illustrate the production of fat from carbohydrate. Yeast, as Pasteur and others have shown, contains at least 1 to 2 per cent. of fat. Its growth in a medium containing no fat, and with sugar as the only possible source of it, which is the case when the nitrate of ammonium is substituted for the tartrate of Pasteur's liquid, affords conclusive proof of the convertibility of carbohydrate into fat.

The fact of the production of fat from carbohydrate is abundantly attested by examples that can be drawn from the higher forms of vegetable organisation, in some of which it is very extensively carried out.

In the oily seeds the fat is preceded by starch and sugar. Sachs* says "Before maturity such seeds contain no fat, but only starch and sugar. Such unripe seeds (*e.g.*, of *Pæonia*) may be detached from the mother plant, and allowed to lie in moist air with the result that the starch disappears and is replaced by fatty oil." Again,† "There is not the slightest doubt that fat is formed in ripening seeds from carbohydrates, particularly starch, since this transformation takes place in the nearly ripe seed, even when taken out of the fruit, when no other material is available under the circumstances for the formation of fat."

If, as appears in the oily seeds, it is susceptible of demonstration that fat is produced from carbohydrates, it is equally demonstrable that carbohydrates are reciprocally producible from fat. The embryo of the oily seed grows in the same manner, and, in its growth, develops the same kind of structure as that of the starchy seed. The cellulose and other carbohydrates found in the seedling are obviously derived from the fatty reserve in the one case, just as they are derived from the starchy reserve in the other.

However inexplicable, from a strictly chemical point of view, such transformations may be, the fact is evident that in the oily seeds fatty matter is, in the first instance, produced from carbohydrate, and, subsequently, in the growth of the seedling, reconverted into carbohydrate.

In the animal kingdom fat is, without question, produced upon a very extensive scale from the carbohydrates. The animal system constitutes, in fact, a laboratory wherein the capacity exists for converting carbohydrate matter into fat. In the milch cow, in the fattening of animals for the table, and in the production of the *foie gras* in the Strasburg goose, we have instances of the extensive operation of the process, and I need not dwell further upon the question of fact, but will proceed to consider that of where and how the change is brought about.

The ferments of the digestive system place the carbohydrate matter

---

* 'Lectures on the Physiology of Plants,' by Julius von Sachs. Translated by H. Marshall Ward, p. 323: Clarendon Press, Oxford, 1887.

† *Loc. cit.*, p. 384.

of our food in a soluble state, if not already existing so. The small intestine is the part of the alimentary tract where absorption of the nutrient matter specially occurs, and, here, the carbohydrate is intimately intermixed with the product of digestion of nitrogenous matter. Thus prepared, the carbohydrate product falls, in the process of absorption, within the sphere of influence of living protoplasm represented by the cells investing the villi. These cells are recognised as the agents concerned in the absorption of fat, but no thought appears to have been given to them as transformers of carbohydrate into fat, although, if we bestow attention to the matter, evidence is seen to be forthcoming suggesting that they, in reality, fulfil this function.

It is well known that after food rich in fatty matter the lacteals are charged with milky chyle, that the cells of the villi are more or less loaded with fat, and that fat globules pass from these cells through the centre of the villus to reach the current in the lacteal system.

Observation conducted upon the vegetable feeder after the ingestion of food rich in carbohydrate matter and poor in fat reveals the existence of a precisely similar state of things. On taking, for instance, a rabbit about four hours after a meal of oats, killing it, and opening the abdomen, coils of the small intestine are seen, especially after a few minutes' exposure, to present a white opaque appearance, with milky streaks or lines upon the surface, due to flow of chyle beneath the peritoneum; and the lacteals of the mesentery, owing to the milky character of their contents, are conspicuously visible. The receptaculum chyli is also, from the same cause, readily perceptible, and, if cut into, gives exit to a strongly milky fluid. On the intestine being laid open, a more or less densely white condition of the internal surface presents itself to view, due to the extent to which the mucous membrane is charged with fat, and the villi stand out as opaque projections.

In order that the condition described may be satisfactorily visible, it is necessary that favourable circumstances should exist. The animal itself must be in a good healthy state. The food must be of a natural kind and sufficiently rich in farinaceous constituents. Moistened oats, in the case of rabbits, have yielded the most marked results. After fasting, with unfavourable food, and in ill-conditioned animals the appearance strikingly differs. The intestine is transparent and watery, and the lacteals are not perceptible. Between this condition and that in which the lacteals are fully injected, any intermediate degree of milky character may, of course, be perceptible.

From the appearances presented, then, to the naked eye, it is learnt that under suitable food, rich in starchy matter, the same passage of fat through the lacteals occurs as after feeding directly with fat. It seems to me impossible that the quantity of fat observed to be thus entering the system could be derived from that contained in a free state in the food. Analysis of the oats consumed in my experiments placed the amount of fat present at 5 per cent., which agrees with the estimations made by others.

Upon the facts before us, the following train of reasoning may, I am of opinion, be legitimately set forth. We know, upon irrefutable grounds, that the capacity of producing fat from carbohydrate matter exists within the animal system, and must be extensively in operation. The intestine constitutes the main seat of preparation of carbohydrate matter for absorption and subsequent utilisation, and from the intestine we find fatty matter flowing into the system through the lacteals, which cannot be adequately accounted for, except on the hypothesis of its origin from the carbohydrate matter of the food. Indisputably, the formation of fat takes place somewhere in the system, and there is no situation more propitiously circumstanced to meet all the requirements of the problem than the intestinal villi. Moreover, with the seat of formation in this position, the introduction of fat into the system as a production from carbohydrate food, is brought into conformity with its direct introduction, preformed, from without. In the one case, the fat is simply absorbed, and reaches the lacteals; in the other, it is produced by assimilation from carbohydrate matter, and then similarly reaches the lacteals.

If we now pursue the matter further, and give attention to the villi, it will be seen that support is afforded to the view that has been advanced.

It may be confidently assumed that it is through the operation of protoplasmic action that the transformation is effected. For many years I have been acquainted with the fact that the lacteals are well charged, as above described, with milky chyle after the ingestion of carbo-hydrate food; but it did not occur to me, until recently, that it was permissible to look to the villi as constituting the seat of the production of fat from carbohydrate matter. I had thought that the most likely position for the transformation was in the liver, and, doubtless, as stands in accord with general opinion, the liver possesses the power of exerting more or less effect in this direction. But, looking to the liver as the special seat of transformation, there came the question of how the fat reached the system. I sought for it in the hepatic vein, and took into account the possibility of its reaching the intestine under the form of the resinoid constituents of the bile. After conducting an extended enquiry into the matter, I failed to obtain evidence reconcilable with the carbohydrate matter of the food being further than to a partial extent disposed of by the agency of the liver. I learnt nothing satisfactory about fat reaching the hepatic vein, and, with respect to the bile, the information obtained failed to sufficiently give the explanation needed.

It was upwards of twenty years ago that my experiments were conducted. Through the medium of a biliary fistula I collected the bile, and estimated the amount of solid matter passing from the liver to the intestine, under various conditions as regards ingesta. The outcome of the results was quite inadequate to account for the fat manifestly produced in the system from the carbohydrates of the food. I saw sufficient to lead me to the opinion that some, if not all, of the fat produced in the liver cells is metamorphosed into resinoid biliary principles, and, in this form, transmitted to the alimentary canal. Here indications presented themselves of the occurrence of re-conversion into, and subsequent absorption by the villi as, fat. The effect of the acid chyme escaping from the stomach is to throw down the glycocholic acid of the bile when the two arrive in contact. An emulsion-like product is formed, from which it has appeared to me that milky chyle has arisen. Further, under certain circumstances, I have witnessed in the strictly fasting dog the lacteals coming from a limited portion of the intestine to be charged with milky chyle, and the interior of this limited portion to contain a coloured bilious fluid, from which other parts were free.

Such was the unsettled position in which the matter formerly stood. A new light has been thrown upon it by the issue of recent enquiry. We look to protoplasm as the agent for effecting the transformation. But protoplasm administers to various offices. In some instances its special purpose is to produce secretions of different kinds, in others it may be designed to undergo metamorphosis into textural elements endowed with particular functional attributes. Assimilative power over food principles may be looked upon as constituting its most primitive capacity, and, as regards this capacity, it is possible that it does not become annulled when other functions are fulfilled. Evidence, at all events, is adducible standing in support of this view.

The protoplasm of the cells of the villi may be regarded as specially intended for the performance of assimilative action. If we study the villi microscopically the appearances seen after the ingestion of food and at a period of fasting are very different. In the fasting state they are covered with a layer of closely set columnar cells provided with nuclei and granular contents. Here and there cells may be discovered containing a few fat particles, but nothing beyond this in the direction of fat is discernible. The appearance presented by the cells is delineated in one of the sketches on p. 253. At a period of digestion numbers of cells become more or less loaded with fat, as shown in the other sketch on the page. Of those which become thus charged, many present a much altered form. The columnar character, it is true, predominates, but the columns are thicker and shorter. Often the cells are conical or pear-shaped. Sometimes they are spheroidal, looking in this state very much like fat-containing cells derived from the liver. Speaking of this resemblance, the cells of the liver, indeed, appear to play a supplementary part in the formation of fat from carbohydrate to that played by the cells of the villi, assimilating the carbohydrate matter which escapes being assimilated in the villi, and which, as a result, is permitted to reach the portal blood. It is interesting to notice that both in the case of the lacteal and the vascular system a second line of protoplasmic matter, consisting of the lymph cells of the absorbent glands, on the one hand, and the cells of the liver on the other, has to be passed before the general circulation is arrived at.

I do not see that any other conclusion is permissible than that in the cells depicted, from villi of the rabbit fed upon oats, the fatty matter was the product of the influence of protoplasm upon the carbohydrate matter ingested. Looked at in their entirety, the points before us stand thus. Beyond all question, as previously stated, the production of fat largely takes place within the animal system from carbohydrate. After carbohydrate food the lacteals are seen to be charged with milky chyle in the same manner as occurs after the direct ingestion of fat. The protoplasmic cells of the villi intervene between the carbohydrate material contained in the alimentary canal and fatty chyle contained in the lacteals. These protoplasmic cells, in the presence of carbohydrate food, are found to become more or less loaded with fat, which finds its way into the lacteal system. Within the villus there is evidently a great protoplasm-generating capacity. Besides the columnar epithelial layer, which, as I have before mentioned, at the time of assimilative activity assumes, to a greater or less extent, an altered character, a number of lymph or wandering cells are to be seen. The two lie in close juxtaposition, and probably both contribute towards rendering the villi richly provided with protoplasm, propitiously placed for exerting an assimilative action on the food principles prepared by digestion for service in the system.

At pp. 254, 255, 256, and 257, photo-engravings of micro-photographs are furnished, representing the villi in section, after fasting and after the ingestion of oats. The intestine was in each case subjected to the usual treatment with osmic acid for staining and bringing into view the fat, and then embedded in paraffin for the section cutting. Preparatory to mounting in Canada balsam the sections were faintly tinted with hæmatoxylin. After fasting, no appearance of fat is presented, whilst, after the ingestion of oats, not only is black stained fat visible in parts of the epithelial layer, but likewise coursing along in the centre of the villus, just as is perceptible after the direct ingestion of fat. Thus the channel of actual entry of fat into the system is shown to be the same whether the fat is directly ingested or, on the other hand, produced from carbohydrate food. In the one case the protoplasm of the villi takes the fat preformed; in the other it forms it from carbohydrate. In both, the channel of entry is through the lacteals.

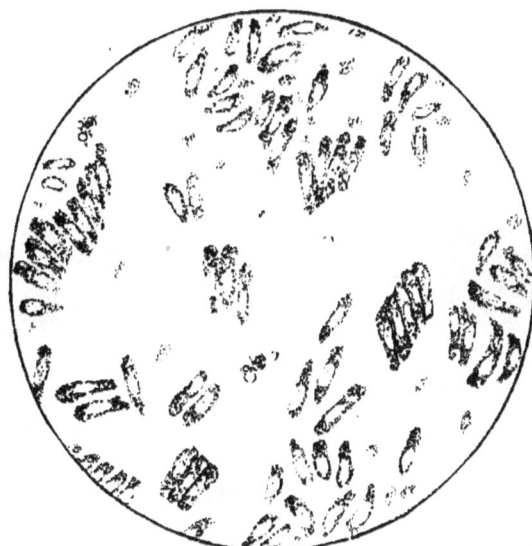

Photo-engraving from sketch showing appearances presented by cells of villi from a fasting rabbit.

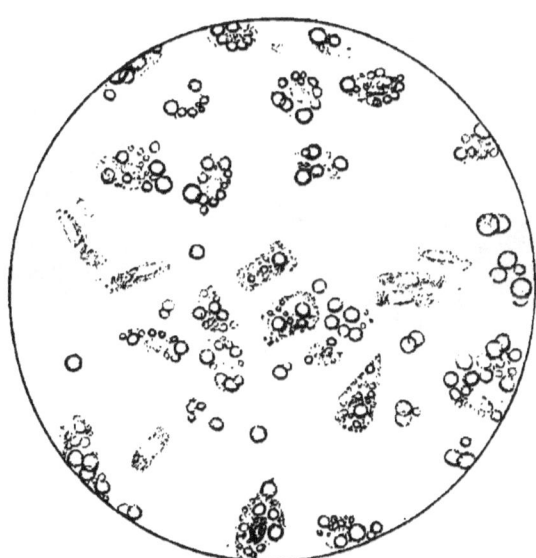

Photo-engraving from sketch showing appearances presented by cells of villi from a rabbit killed four hours after having been fed with oats.

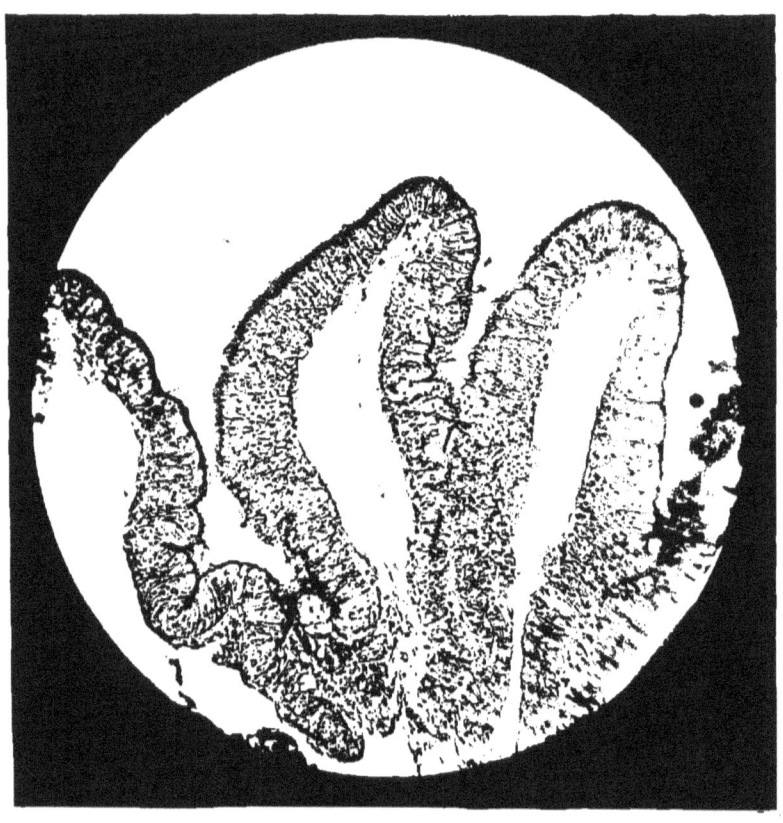

Villi in section, magnified 100 diameters, from fasting rabbit. Treated with osmic acid. No fat perceptible.

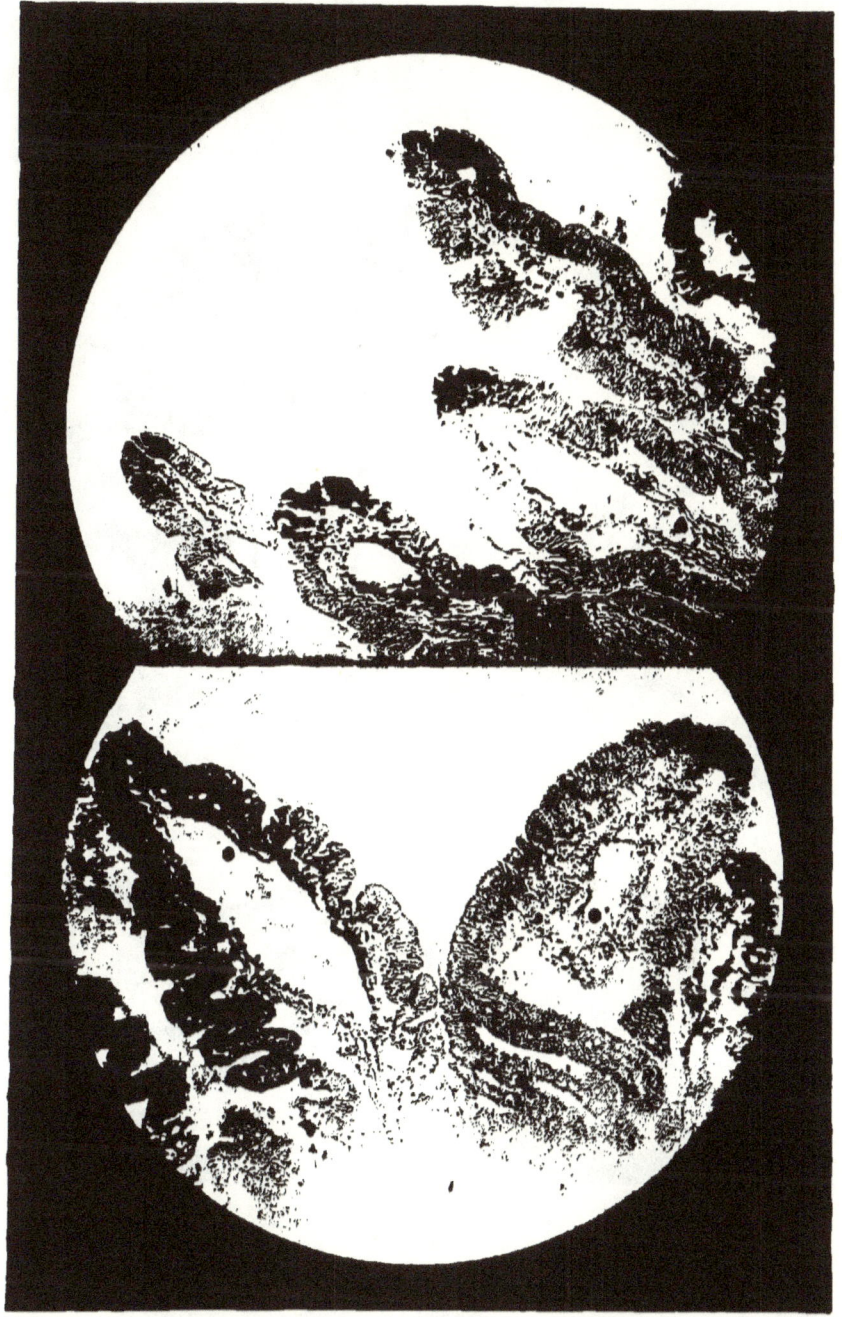

Villi in section, magnified 100 diameters, from rabbit killed four hours after having been fed with oats. Treated with osmic acid. Blackened fat largely visible.

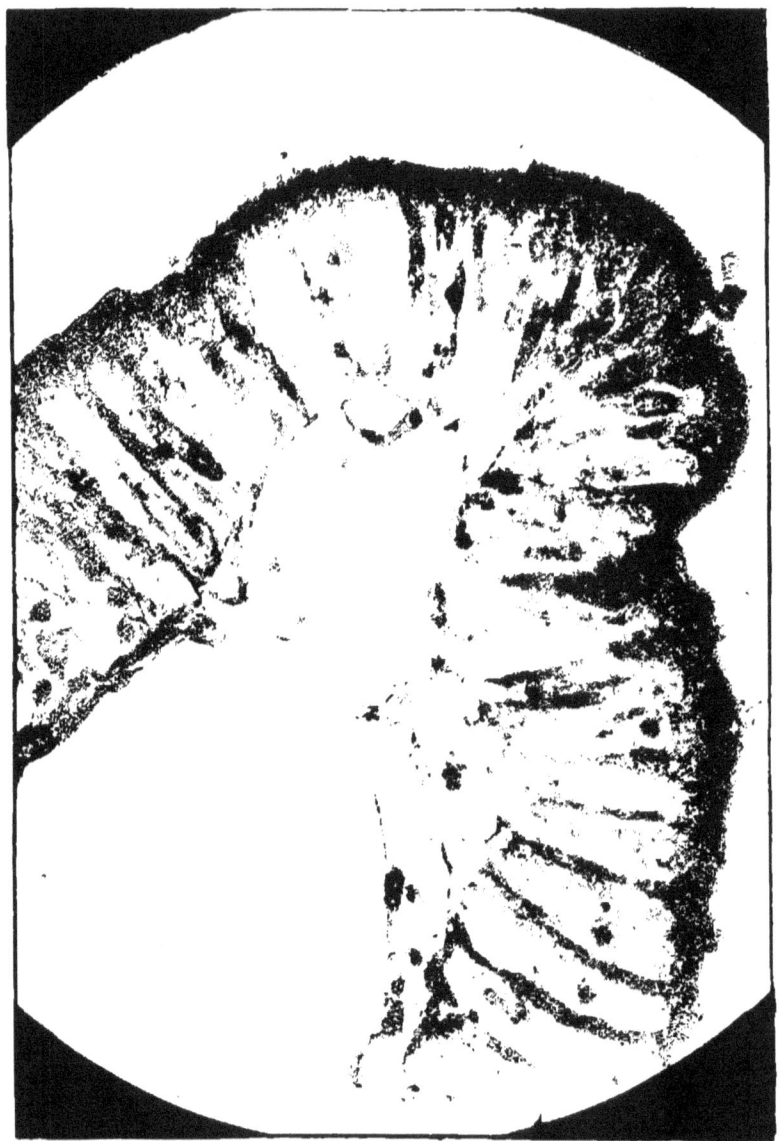

Point of villus in section, magnified 600 diameters, from fasting rabbit. Taken from a portion of the section represented at p. 254. No indication of fat.

Point of villus in section, magnified 600 diameters, from rabbit killed four hours after having been fed with oats. Taken from a portion of the section represented at p. 255. Blackened fat displayed.

The cells of the liver, as I have intimated, probably supplement those of the villi in producing fat from the carbohydrate which escapes disposal in the villi and reaches the portal vein as sugar. If this be the case, we ought to find in the process of fattening with carbohydrate food that the liver gives evidence of being involved to a greater extent in the operation in question, where the villi and lacteal system are less perfectly developed than under the opposite condition. Now, this happens to accord with what is actually observed. In the bird the villi are much less highly developed than they are in the mammal, and in the process of fattening (for the production of the *foie gras*) to which the Strasburg geese are subjected the liver, as is known, attains enormous size from the deposition of fat. In the mammal the case stands otherwise. Many years ago I visited the slaughter-houses to which some of the fat animals exhibited at the Christmas London Cattle show were taken, and was struck with the healthy and fleshy-looking state of the liver and the other abdominal organs. With efficiency on the part of the villi, the liver would escape the tax of work that under other circumstances would be thrown upon it.

I have for a long time held that what in the human subject is spoken of as fatty degeneration of the liver is in reality the result of conditions leading to excessive functional production of fat. Conditions may exist attended with the ingestion of carbohydrate matter beyond the capacity of the villi for disposing of it to the usual extent, and thus leave more than the ordinary amount for undergoing conversion into fat by the agency of the liver.

As regards the mode of transformation, by protoplasmic action, of carbohydrate into fat, I consider that it may be reasonably surmised that the process is not a direct one, but one in which the production of proteid plays an intermediate part. It has been pointed out (p. 240) that in the transmutation of carbohydrate matter to a lower form of hydration in the vegetable organism, as, for instance, in the deposition of fabric cellulose and storage starch, the weight of evidence and probability is in the direction of its being effected by entering, in the first instance, into the constitution of the proteid of the protoplasm that is instrumental in bringing about the change, and then being deposited in the altered form. Similarly, the protoplasmic matter of the cells of the villi may first lead to the incorporation of the carbohydrate into proteid, from which, by cleavage, the fat may be afterwards thrown off.

If we here pause for a moment to look at what is observed to occur elsewhere, it will be seen that there is nothing irrational in the view that has been put forward.

According to the description that is given in physiological works of the secretion of milk by the mammary gland, the alveoli are lined with cells consisting of growing protoplasm, which undergoes metabolic change so as to give rise to the appearance within its substance of fat globules of various sizes that become discharged, and constitute the fatty portion of the milk. Protoplasm exists to begin with, and from this the elements of the milk are derived apparently, it may be stated, by a process of splitting up. The cleavage of proteid may give rise not only to the fat but likewise to the lactin and casein, and in support of casein constituting a product, from the antecedent of which carbohydrate has been cleaved, we have the fact, as I mentioned (p. 32) when speaking of the glucoside constitution of proteid matter, that only an insignificant amount of cleavage carbohydrate is obtainable from it by chemical means, compared with what is obtainable from albumin and other proteids.

Again, the same kind of description is employed for representing the deposition of fat in the development of the adipose tissue of the body. The deposition takes place in the cells belonging to the connective tissue. These multiply and become charged with oil globules, which grow larger and coalesce, until ultimately the cell consists of one large spherical oil globule, the protoplasmic remains of the cell forming a thin capsule around it. The protoplasm of the cell thus visibly becomes replaced by fat, which we are driven to look upon as split off from the pre-existing proteid substance.

In the peculiar disease known as myxœdema, which is attended with an undue accumulation of connective tissue conspicuously infiltrated with a redundance of its proteid accompaniment—mucin, the idea, in view of the doctrine I am advocating, forces itself upon the mind that the condition may be due to an imperfect performance of proteid cleavage and liberation of fat. The province of protoplasm is to construct, whilst the effect of ferment action is to split up. The redundance of connective tissue and mucin may be referable to default existing in connection with the latter process. It may be, in the disease, that there is a want or deficiency of the proper ferment for splitting off fat from proteid, and that the curative effect of the thyroid extract treatment is attributable to the want being met by introducing into the system the ferment derived from an external source.

In the fatty degeneration of muscle it may further be taken that an allied cleavage of fat from proteid occurs. Although the special object of the protoplasm may not be, as in the instances that have been dealt with, to effect the formation of fat, the proteid— indeed, the same may be said of tissue proteid anywhere existing— may undergo dissociation with the liberation of fat when conditions deviating from those which tend to the maintenance of a state of integrity prevail.

From the train of reasoning before adduced (p. 245) the conclusion is deducible that the actual source of all fat is the primordial carbohydrate produced by the chlorophyll corpuscles of the vegetable kingdom actuated by the power derived from the sun. Starting with this carbohydrate, the operation of living protoplasm is needed for its conversion into fat, and the first step of the process presumptively consists of incorporation into proteid. The step to follow will be one of cleavage, and this may ensue either at once, as, for example, in the case of the villi, or subsequently, as in the case of the mammary gland, the growth of adipose tissue, and the fatty degeneration of muscular and other structures.

There may be, it is true, changes taking place within the living system of an ordinary chemical nature, but the two mainsprings of power determining the chemical metamorphoses of life are protoplasmic and ferment actions. The effects produced by these are directly opposed. Whilst we witness as the result of the operation of the one a synthesising or constructive effect, the effect of the other is to dissever or split up. It is, doubtless, upon a delicately-adjusted balance of these two operations that the play of changes belonging to life depends. Assuming carbohydrate to have been appropriated by the instrumentality of protoplasmic action to the construction of proteid, we have a body to deal with from which either carbohydrate or fat may be subsequently evolved according to the surrounding determining conditions. The blood supply (embracing the state of the blood which is influenced by that of the arteries through their muscular coat governed by the nervous system) probably constitutes a great factor in the determination of the result. At all events, considerations, it will be seen, can be adduced, giving support to this conjecture.

Observation, for instance, shows that the conditions conducive to fat deposition are sluggish circulation, deficient blood supply, and deficiency of the red corpuscular element of the blood, all of which tend to produce a deficiently-oxygenated state. The fattener of animals knows that, with a given amount of fattening food, his object is more speedily attained by keeping the animal in a state of quiescence, and, further, in a darkened place, the effect of which is to favour escape from the temporary excitations of the vascular system produced by nerve action. The more vegetative the life of the animal, to a greater extent does it become a fruitful fat producer. Both functional fat production from carbohydrate and fat cleavage from tissue proteid are favoured.

In illustration of the effect of deficient blood supply, and also of deficiency of the red corpuscular element of the blood, in promoting the cleavage of fat from proteid, I may refer to the fatty degeneration of the muscular fibres of the heart which is observed as an accompaniment of ossification and other obstructive conditions affecting the coronary arteries, and also of the morbid state known as idiopathic or pernicious anæmia. The occurrence of fatty degeneration—proteid fat liberation—it may be added, under the circumstances named, is nothing more than a phenomenon standing in harmony with a recognised pathological principle applicable to the tissues generally.

In advanced diabetes, proteid cleavage with the liberation of sugar, instead of fat as in fatty degeneration, occurs, and observation shows that an opposite kind of blood influence prevails.

In mild forms of the affection, and in the early stage of the grave class of case, the tissues do not suffer when the disease is held under by appropriate dietetic management. Under these circumstances, the only fault existing is a defective application or assimilation of the carbohydrate matter ingested. That this is the case is proved by the train of reasoning I will proceed to set forth.

The question presented for solution is whether protoplasmic assimilative action is at fault, or whether assimilated carbohydrate is brought back into sugar by the operation of undue ferment influence. As the tissues, under the circumstances named, supply no evidence of being implicated, the exercise of the undue ferment influence, if it exist, must be located in the liver; and, as is known, we have here transmuted sugar under the form of glycogen to deal with which, if re-transformed into sugar by ferment action, would account for the phenomenon we are considering. Now, should the result be attributable to ferment action, sugar ought to be discharged with the urine after the ingestion of limited as well as of larger amounts of carbohydrate. The ferment would operate alike in each case, and lead to sugar production equal to the extent of its power of action. But, in the cases in question, carbohydrate can be ingested up to a certain amount (the amount varying with, and remaining steady in, the individual case) without being attended with the elimination of sugar. There is the capacity of disposing of ingested carbohydrate matter within the system up to a certain point without leading to its running off as sugar. The only difference between such a case and the healthy state is that the capacity is more limited in the one than it is in the other, and we are driven to conclude that the fault must consist of an impairment of the power possessed by the system of appropriating the carbohydrate, and preventing its escape from the body as sugar.

I hold, as enunciated in this work, that protoplasmic action is the agency by which the appropriation of carbohydrate matter is effected, and that the agents of appropriation are the cells of the villi and of the liver. Under normal circumstances, carbohydrate is not allowed to pass these lines of stoppage. Conditions leading to its doing so occasion saccharine urine, which stands proportionate in extent to the amount of carbohydrate that escapes arrest, and is thus permitted to reach the general circulation. All the facts disclosed by experience bearing upon diabetes stand in conformity with this assertion.

I have said that we must look to deficient protoplasmic action as the source of the sugar which passes off as waste material from ingested carbohydrate. It cannot be conceived, however, that the error is due to the protoplasm itself. This, it may be considered, is endowed with an inherent power which, as long as it is in a living state, it retains. For the due exercise, however, of this power proper surrounding conditions are needed. The state of the blood I regard as a most important factor in relation to this matter. An unduly oxygenated state can be demonstrated to constitute a source of saccharine urine; and, it is interesting to note, that carbonic oxide and also, it is to be said, the nitrites—agents which, in each case, produce analogous compounds with hæmoglobin to the oxygen compound, oxyhæmoglobin—lead to the same result. Now, enough evidence exists to show that an unduly oxygenated condition of the blood, arising from vasomotor paralysis, and explicable by the increased transit through the capillaries failing to permit de-arterialisation to the ordinary extent to occur, actually exists in connection with diabetes.

A vasomotor paralysis implicating only the vessels of the chylopoietic viscera may stand at the foundation of the form of diabetes limited to defective assimilation of ingested carbohydrate. The red, or raw meat, appearance of the tongue, and it may be of the whole interior of the mouth, including the fauces and pharynx, that is sometimes seen, affords evidence of vasomotor paralysis of a more extended kind, and in harmony constitutes, as experience shows, an indication of severity. In cases occurring in persons of advanced years, the disposition exists for the complaint to be confined to a defective appropriation of ingested carbohydrate. In cases, on the other hand, occurring amongst young persons, whilst, as a rule, defective appropriation of ingested carbohydrate constitutes the only condition existing at first, the affection progresses, and, in the course of time, sugar is eliminated beyond what can be accounted for by the food, and is evidently in part derived from the tissues of the body. The condition which at one time only leads to a manifestation of defective assimilation subsequently, on attaining a more advanced stage, gives rise to dissolution of constructed proteid with the cleavage of sugar, and thus produces the more intensified form of disease that becomes developed.

It will be observed from what has been said that the cleavage of fat and the cleavage of sugar from tissue proteids are associated with opposite states of the contents of the vascular system. An under-oxygenated state is noticed in association with the one, an over-oxygenated state in association with the other. Probably the processes of fat and carbohydrate cleavage are always being carried on to a certain extent, the carbohydrate cleavage resulting under natural circumstances in glycogen and not sugar, and accounting for the glycogen met with in various textures.

Reading nature by the light of the evidence that has been set forth, we learn that, as a result of the operations of life, carbohydrate matter becomes (1) transmuted to a lower state of hydration, (2) applied to the production of proteid, and (3) transformed into fat. Some other changes of minor import contributing to subsidiary offices in the economy may take place, but I have no misgiving about carbohydrate matter passing, in the main, in the directions named for application to the purposes of life. Observers have hitherto been looking for the acquirement of knowledge concerning the manner in which carbohydrate matter becomes disposed of in the animal system by the prosecution of research in the opposite line of inquiry. Fruitless results have attended, and, I am of opinion, are not likely to do otherwise than continue to attend such efforts. From an abyss of darkness, constituting the outcome of research conducted in the direction of change produced by ferment action and chemical agents, we are led, by research applied in the direction of protoplasmic action, to a clear and intelligible position. Harmony exists throughout, and the whole thing can be epitomised so as to lie in a nutshell. By protoplasmic action carbohydrate matter, as observation teaches us, is disposed of in a certain manner. If not disposed of in this manner, it escapes from the system as useless material, producing the condition belonging to diabetes.

INDEX.

	PAGE
Absorption of sugar from stomach	83
Achroodextrin	11
Acrose, α, an artificially synthesised six carbon atom sugar	5
Acid, citric, cane-sugar inverting power of, at body temperature	98
—— —— employment of, for inversion of cane-sugar	92—93, 108
—— glycuronic, asserted presence in urine after chloroform	163
—— hydrochloric, cane-sugar inverting power of, at body temperature	98
—— lactic, probable formation of, from glucose in alimentary canal	100
—— —— probable formation of, from lactose in alimentary canal	99
—— sulphuric, action of, on carbohydrates	92
—— —— relative effects of 2 per cent. and 10 per cent. strengths upon cleavage carbohydrate	213
Acids, action of, on cane-sugar	92, 98
—— effect on action of amylolytic ferments	83, 85—87
Albumin, egg, cleavage experiments on	31
—— —— preparation and purification for potash cleavage experiments	33—34
—— —— preparation and purification for treatment to obtain cleavage sugar with sulphuric acid	44
Alcohol-coagulation does not destroy sugar-producing ferment of liver	147—154
Alcohol, employment in process of glycogen (amylose carbohydrate) extraction and estimation	63
—— strength required for precipitation of proteid-cleavage product	33
—— use of, in extracting sugar	58—62
Ammonia, interference of, with employment of Fehling's solution	178
—— principle of action of, in the ammoniated cupric test	71—80
—— vitiating influence of, in gravimetric sugar determinations	70
Ammoniated cupric test equally applicable to fresh and decomposing products	173
—— —— —— mode of titration with	75—78
—— —— —— precautions in use of	72, 77—78
—— —— —— precision attainable with	78—80
—— —— —— quantitative determination of sugar by	71—80
—— —— —— reliability shown by close accord in duplicate analyses and with the results yielded by the gravimetric process	79
—— —— —— solution, composition and preparation of	71—74
—— —— —— —— self-preservative power of	71
—— —— —— —— sugar value of	73—74
—— —— —— —— standardisation of	74
Amygdalin, example of nitrogenous glucoside	27
Amylodextrin	11

	PAGE
Amylodextrin, constitution of	45
Amyloid substance, employment of term	212
Amylose carbohydrate, employment of term	213
—— —— in blood and tissues	215—220
Amyloses, general description of	6—12
Analytical procedure for recognition and estimation of carbohydrates in animal products	58—80
Animal, cold-blooded, sugar in liver of	140—143
—— charcoal, employment of, for decolorising saccharine products	180
—— food as a source of glycogen	118—121
—— —— both free and cleavage sugar derived from	118—121
—— —— sugar in portal blood after	104
—— gum	10
—— —— resemblance of proteid-cleavage product to	35, 41
—— —— the carbohydrate from the glucoside mucin	27
Apparatus for sugar titration, description and figure of	75—76
Arabinose, a five carbon atom sugar	1
Asparagin, conjugation with carbohydrate to form proteid	239
—— participation in synthesis of proteid in plant	54—55
Autoclave, employment for aqueous extraction of glycogen from liver	124
—— employment in analysis	67
Beef tea, presence of sugar in	118—119
Bernard, alleged tolerating capacity of blood for sugar	187—193
—— discovery of glycogen by	211
—— glycogenic theory propounded by	111, 112, 113
—— on amount of sugar in blood	162—163
—— on disappearance of sugar from drawn blood	171
—— on influence of body temperature on amount of sugar in liver after death	141
—— statement about sugar in arterial and venous blood	166
Berthelot on artificial glucosides	53
Bile, fat from resinoid matter of	250
—— influence of, in digestion	84
Blood after withdrawal, in relation to sugar	171
—— alleged tolerating capacity for sugar	187—193
—— amount of sugar in	158—165
—— analysis for sugar	159, 161
—— and tissues in relation to amylose carbohydrate	215—220
—— comparison of arterial and venous, in relation to sugar	166—171
—— evidence given by, of sugar-arresting action of liver	109—112
—— examination of, for carbohydrate	59—64
—— facts connected with, refute glycogenic doctrine	223—225
—— hepatic venous, condition of, in relation to sugar	110—111
—— increases sugar production in washed liver	147
—— influence of, on sugar production in liver after death	137
—— in relation to sugar	157—177
—— Lépine's theory of glycolysis in	176—177
—— nature of sugar in	157—158

INDEX.

	PAGE
Blood of general circulation, amount of sugar in......................	161
—— —— —— not variable as portal, in relation to food	112
—— —— —— standard amount of sugar in	101
—— —— —— uninfluenced by ingestion of carbohydrate...............	163
—— of hepatic veins, saccharine state found after death one of main supports of glycogenic theory	136
—— portal, after animal food gives evidence of absorption of sugar	120—121
—— —— amount of sugar in	101—108
—— —— and general, relative amount of sugar in......................	111
—— —— carbohydrates from ingestion traced to	81
—— —— nature of sugar in	158
—— —— product of starch digestion in	89—90
—— —— supports proteid formation from carbohydrate	242
—— —— the carrier of ingested carbohydrate to the liver	117
—— —— variable condition of, as regards sugar in relation to food not transmitted to blood of general circulation	112
—— *post-mortem* not to be taken, as regards sugar, as representative of *ante-mortem* state...	160—161
—— precautions requisite in collection for analysis	110, 160, 163—165
—— state of, influencing fat and sugar production....................	260—261
Boiling, method for preventing *post-mortem* ferment action in liver	134—135
Brain, sugar in	204
Brown and Morris on the dextrins.....................................	11
Brücke's process for separation of sugar from normal urine............	179—180
Brunner, glands of..	87
Cane sugar, absence of cupric oxide reducing power	92—93
—— —— an illustration of dehydrating action	21
—— —— chemistry and properties of	12—13
—— —— citric acid the best agent for inversion for quantitative determination..	67
—— —— constitution of, ...	44
—— —— digestion of..	91—98
—— —— estimation of, in presence of other carbohydrates	93
—— —— form of sugar in portal blood after ingestion of	98—99, 108
—— —— inversion, interfering action of sodium sulphate	108
—— —— —— of, by acids..	92, 98
—— —— inverted by dilute acids at body temperature...................	98
—— —— —— by ferment in stomach (to small extent) and intestinal walls	94—96
—— —— portal blood after ingestion of............................	107—108
—— —— relation to optical rotation	12—13
—— —— uninfluenced by salivary and pancreatic secretions	93
—— —— use of, in standardising ammoniated cupric test	74—75
Carbohydrate, amylose, analytical procedure for examination of animal products for	63—64
—— and fat mutually convertible	247
—— application to proteid formation	233, 239—245
—— chemical constitution of..	5
—— dehydration by protoplasmic action	3

INDEX.

	PAGE
Carbohydrate, deposition from proteid in plant	55—57
—— destination of, in animal system	232—264
—— effects of protoplasmic action upon	233
—— food a source of fat	248
—— —— sugar in portal blood after	81, 105—108
—— genesis of	1—4
—— ingested, a source of glycogen	113—117
—— incorporation in construction of proteid	53—55
—— mode of transformation into fat by protoplasmic action	258
—— primordial source of all carbon compounds of nature	245
—— reserve of plants, an illustration of dehydrating action	21
—— transmutations in living plant	25—26
—— universal distribution of	233
—— used in formation of proteid and fat by cells of villi	121
Carbohydrates, artificial synthesis of	5
—— chemical characters and relations of	5—16
—— classification of	6
—— elementary composition of	1
—— general formula for	1
—— how naturally disposed of in the animal system	221—264
—— ingested, liver in relation to sugar from	109—131
—— —— portal blood in relation to	101—108
—— sense in which term used	1
—— stopped from reaching general circulation by cells of villi and liver	262
—— synthetic formation in plant	2
—— transmutation of	18—26
—— —— —— in relation to operations of life	22—26
—— —— —— with decreased hydration	19—22
—— —— —— with increased hydration	18—19
—— varied number of carbon atoms in molecule of	1
—— vegetable kingdom the source of	1
Carbonic acid, source of carbon of carbohydrate	2
—— oxide inhalation leads to production of glycosuria	145
Casein, a suggested cleavage-product from proteid	32, 245
—— cleavage experiments on	32
Cells of liver, fat formers	251
—— of villi, fat formers	251
—— —— —— photo-engraving of, showing appearance of, after fasting	253
—— —— —— —— —— after ingestion of oats	253
Cellulose, behaviour of, with sulphuric acid	7
—— chemistry and properties of	6—8
—— deposition in plant	3—4
—— dissolving power of fungi upon	91
—— formation by yeast cells	3
—— in relation to digestion	90—91
—— in yeast, an illustration of dehydrating action	21
—— resistance of, to solvents	90—91
—— ultimate transmutation product in plant	234
Chlorophyll corpuscles, formation of starch by	55—56

INDEX.

	PAGE
Chlorophyll corpuscles, seat of primary carbohydrate formation	4
—— power of fixing carbon	2
Chyle, milky, after carbohydrate food	248
Chyme, neutralisation of, by alkali of bile, &c.	84
Citric acid, effect of, on liver-ferment	155
—— —— use of, in inversion of cane-sugar for quantitative determination	74
Cleavage-carbohydrate and glycogen from various structures	211—220
—— —— effects of different strengths of sulphuric acid upon	213
—— of proteid, origin of carbohydrate by	3—4
—— sugar from digestion of animal food	119—120
Cold-blooded animal, sugar in liver of	140—143
Colloidal properties of starch	81
Conversion of amyloses and saccharoses into glucose by sulphuric acid, time required for	66—67
Copper compound from potash proteid-cleavage product	35, 42—43
Cupric oxide reducing power not possessed by cane sugar	92—93
—— —— —— —— of sugar of portal blood	102

Decomposition, effect upon disappearance of sugar from drawn blood	173, 176
Dehydration changes the result of action of living matter	18, 19—22
Dextrin, chemistry and properties of	10—12
—— example of artificially induced dehydration	20
—— probable nature of action giving rise to	10—11
—— production of, in pancreatic digestion of starch	84
—— source of	10—11
Dextrins, production of, in process of starch digestion	82
Dextrose, chemistry and properties of	15—16
—— convertible into lævulose	52
—— different optical varieties of	52
—— sources of	15
Diabetes, a faulty disposal of carbohydrate matter	222
—— association with cleavage of sugar from proteid	229
—— due to defective assimilation of carbohydrate	229
—— glucose the form of sugar eliminated after ingestion of lactose	99
—— inhibitory action of liver	112
—— in relation to effects of ferment and protoplasmic actions	262
—— mild form, defective assimilation of carbohydrate only	261
—— —— —— toleration to some extent of carbohydrate food	262
—— production of sugar from tissues in	263
—— rationale of dietetic management of	228—232
—— relation of sugar in urine to that in blood	191—193
—— —— to vaso-motor paralysis	263
—— severe form, sugar in part derived from proteid-cleavage	261
—— teachings of, refute glycogenic doctrine and point to mode of disposal of carbohydrate matter	227—232
Diastase, example of unorganised ferment (amylolytic)	19
Diet, regulation of, in diabetes	228—232
Digestion, cleavage-sugar from proteid matter produced in	119—120
—— of cane sugar	91—99

	PAGE
Digestion of cellulose	90—91
—— of lactose	99
—— of starch	81—90
—— —— —— necessary for its absorption	82
Egg, sugar in	206—210
Enzyme, see ferments, unorganized.	
Erythrodextrin	11
Erythrose, a 4-carbon atom sugar	1
Fasting, sugar in portal blood after	103
Fat and carbohydrate mutually convertible	247
—— as seen in villi in section after ingestion of oats, photo-engravings of micro-photographs	255, 257
—— derived from carbohydrate food	348
—— formation from carbohydrate	233, 245
—— —— —— —— by cells of villi	121
—— —— —— —— in oily seeds	247
—— —— in villi from carbohydrates of food	248
—— from carbohydrate by protoplasmic cells of villi	252
—— mode of production from carbohydrate by protoplasmic action	258
—— of adipose tissue, cleavage from proteid	259
—— of milk, a cleavage product from proteid	245
—— origin in yeast from carbohydrate	246
—— production of, from carbohydrate by cells of liver	258
Fatty degeneration of liver a result of excessive fat formation	258
—— —— of muscle fat cleavage from proteid	260
Fehling's solution, composition of	68
—— —— author's modification of	68
Ferment action attended with increase of hydration	18—19
—— —— influence upon cellulose	19
—— —— in relation to operations of life	22—24
—— —— nature of	18—19
—— —— proteid-cleavage sugar produced by	49—51
—— amylolytic, in certain digestive secretions	82
—— and protoplasmic actions opposed in their effects	260
—— —— —— —— reciprocal play of	24—26
—— cane sugar inverting, not destroyed by alcohol	95—96
—— —— —— —— contained in stomach (to small extent) and intestinal walls	94—96
—— glucose-forming, in intestine	87—89
—— of liver, glucose-forming capacity of	146
—— pancreatic, experiments with	84
—— (sugar-forming) of liver	133—140
—— —— —— —— analogous position to fibrin ferment of blood	146
Ferments, action of, on cellulose	90—91
—— amylolytic	19, 22
—— peptonising or proteolytic, in animal and in plant	22—23
—— production in seeds and buds	25—26

INDEX.

	PAGE
Ferments, unorganised	19
Fermentability, diversity presented by the various sugars	52
Fermentation, acetic acid	18
—— alcoholic	18
—— distinct from ferment action	18—19
—— lactic acid	18
—— occurrence with sugar of normal urine	181
—— negative behaviour of proteid-cleavage-sugar	40—41
—— putrefactive	18
Fibrin, cleavage experiments on	32
Filter-paper, danger of using, in dealing with carbohydrates	63, 214
Fischer, E., on optical activity of substances	41
—— —— varieties of sugar	52
—— on recovering sugar from an osazone	46, 49
—— on the phenylhydrazine reaction of sugars	17
—— work on synthesis of sugars	5
Foie gras, result of fat formation from carbohydrate food	247, 258
Food, carbohydrate, influence on amount of glycogen in liver	127
Freezing, method of checking *post mortem* transformation of glycogen	125, 128, 129, 134—140
Frog, illustrating influence of body-temperature at death on amount of sugar in liver	141
—— liver rich in glycogen	131
Fungi, cellulose-dissolving power of	91
—— occurrence of glycogen in	9—10, 235
Galactose, chemistry and properties of	16
—— production from lactose	16
Gastric juice, effect of, on starch digestion	83
Gelatin, cleavage experiments on	32
Generative organs of Fish and Crustacea, sugar in	205
Genesis of carbohydrate	1—4
Glass-wool, use of, for filtration	62, 63—64
Glucosan, example of artificially induced dehydration	20
—— formation from artificial glucosides	53
—— —— —— dextrose	16
Glucosazone, crystalline character of	17
—— from sugar of liver taken in ordinary way after death, photo-engraving of	140
Glucose, carbohydrate matter estimated as	127, 148
—— escape with urine after injection into circulation	187—190
—— form in which ingested cane sugar reaches portal blood	99, 108
—— -forming capacity of liver-ferment	146
—— -forming ferment of intestine	87—89
—— form of sugar in liver after *post-mortem* change, but not previously	138
—— —— —— in blood of general circulation	157
—— found in urine after excessive ingestion of cane sugar	116
—— in urine of diabetic after ingestion of lactose	99
—— not changed by digestion	100

	PAGE
Glucose not to any large extent produced from starch within alimentary canal	89
—— subsidiary production of, by action of saliva on starch	82
—— the kind of sugar in the egg	206—207
—— the sugar usually eliminated in diabetes	239
Glucoses, general description of	15—16
—— optically inactive, in sugar cane	41
Glucosone-like body obtained in process for recovering cleavage sugar from its osazone	47
Glucoside constitution of proteid matter	27—57
—— —— —— —— bearings of, in relation to diabetes	229
—— —— —— —— physiological significance of	240
—— illustration of cleavage	3
Glucosides	27—57
—— artificial	53
—— physiological interest and importance of	28
Gluten from wheat flour, cleavage experiments on	31
Glycerose, a 3-carbon atom sugar	1
Glycogen, amount of, in liver after animal food	114
—— —— —— —— after diet rich in carbohydrates	114—117
—— —— —— —— under different conditions and in different animals	125—131
—— and cleavage-carbohydrate, process for estimation of	211—214
—— and proteid-cleavage carbohydrate from various structures	211—220
—— animal food as a source of	118—121
—— a transmutation product in animal	235
—— chemistry and properties of	9—10
—— difficulty of complete extraction by water from liver	122—125
—— diminution compared with sugar production in liver-ferment experiments	148
—— discovery of	113, 211
—— estimated in analysis as glucose	127
—— existence in yeast and other fungi	9—10
—— formation by yeast cells	3
—— general relations as a constituent of liver	122—131
—— ingested carbohydrate matter a source of	113
—— in liver an illustration of dehydrating action	22
—— in liver from excess of carbohydrate food	244
—— in yeast an illustration of dehydrating action	21
—— old method of estimation of	114
—— physiologically a misnomer	9
—— *post-mortem* transformation of, in liver	128
—— process for extraction and estimation of	28, 63—64
—— rapid *post-mortem* transformation of, in liver	125—126
—— recognition in liver-cells	122
—— representative of starch in animal kingdom	81
—— the term physiologically a misnomer	212
Glycogenic doctrine, refutation of	221—227
—— theory (Bernard's), fallacy at the foundation of	111
—— —— initiation of	112—113

	PAGE
Glycolose, a 2-carbon atom sugar	1
Glycolysis in blood, Lépine's theory	176—177
Grape sugar, see Dextrose.	
Gravimetric process of sugar estimation	69—70
Gummose, sugar from Landwehr's animal gum	27
Horse, tissues rich in glycogen	130—131
Hydrazones	17
Hydrolytic action of acids and ferments	18—19
Incubation, effect on sugar of egg	207—210
Intestinal juice, see succus entericus	87
Intestine, cane sugar inverting power possessed by	94—96
—— —— —— —— exceptional state in ruminant	94, 96
—— of ruminant, small cane sugar inverting power of	94, 96
Inversion of cane-sugar by acids	92, 98
Invertin, existence in yeast	21
—— example of unorganised ferment	19
Invert sugar, composed of dextrose and lævulose	13
—— —— formation from cane sugar	13, 92
Kidney, sugar belonging to	201
—— sugar elimination by	162
Lacteals, milky injection of, after carbohydrate food	248
Lactose, action of acetic and citric acids on	13—15
—— chemistry and properties of	13—15
—— constitution of	44
—— conversion by sulphuric acid into dextrose and galactose	15
—— cupric oxide reducing power of	13
—— digestion of	99
—— glucose in urine of diabetic after ingestion of	99
—— osazones from, under varied conditions	14
—— possibly a proteid-cleavage product	32, 245
—— probably partly converted into lactic acid in alimentary canal	99
—— time required for conversion effected by sulphuric acid	67
Lævulosan, example of artificially-induced dehydration	20
—— formation from lævulose	16
Lævulose, chemistry and properties of	16
—— different optical varieties of	52
—— escape with urine after injection into circulation	190—191
—— sources of	16
Landwehr, on mucin and animal gum	27
Landwehr's " animal gum," copper compound of	35
Lieberkühn, glands of, function of	87
Liebig's classification of foods as regards carbohydrates untenable	244
Light, influence on carbohydrate formation	4
Liver, amount and nature of sugar in, at moment of death and afterwards	136—139

T

	PAGE
Liver an agent for carbohydrate transmutation	235
—— cells supplement cells of villi in production of fat from carbohydrate	258
—— fat formation in	249
Liver-ferment, observations on, after alcohol coagulation	150, 154—156
—— possesses a glucose-forming capacity	146
Liver, in different animals, in relation to glycogen	127—131
—— ingested carbohydrate, a source of the glycogen in	113—117
—— in relation to sugar	132—156
—— in relation to sugar from ingested carbohydrates	109—131
—— in same position as other structures in relation to amount of sugar present	144
—— general relations of glycogen in	122—131
—— not, as asserted, the seat of production of sugar from proteid of animal food	121
—— of cold-blooded animal, amount and nature of sugar in	140—143
—— *post-mortem* production of sugar in	133—140
—— production of fat by cells of	258
—— saccharine state found after death one of main supports of Glycogenic Theory	136
—— so-called fatty degeneration of, a result of excessive fat formation	258
—— sugar-arresting function of: evidence of afforded by the blood	109—112
—— —— —— —— —— evidence of afforded by the organ itself	112—118
—— sugar-forming ferment of	133—140
Lung, sugar in	202—203
Liver-substance, sugar production in, after coagulation by alcohol	147—154
—— true state of, stands against glycogenic doctrine	223
—— washed free from blood, sugar production in	146—147
Living matter, special power possessed by	24
Maltodextrin	11
—— constitution of	45
Maltose, chemistry and properties of	12
—— constitution of	44
—— conversion by sulphuric acid into dextrose	12
—— main end product of action of saliva on starch	82
—— portal blood after ingestion of	107
—— production of, in pancreatic digestion of starch	84
—— product of starch transformation	11, 12
—— time required for conversion by sulphuric acid	66—67
Melting points of cleavage sugar osazone, galactosazone, and glucosazone	48
Milk sugar, see lactose.	
—— the fat, lactin, and casein of, cleavage products from proteid	259
Mucin, "animal gum" the cleavage product of	35
—— cleavage experiments on	32
—— example of animal glucoside	27
Mucous membrane, intestinal, glucose-forming power of, even when dried	89
Muscle, sugar in	194—199
Myronic acid, example of sulphur-containing glucoside	27
Myxœdema, probable result of arrest of proteid-cleavage	259

INDEX. 275

	PAGE
Optical activity, diversity presented by the various sugars	52
—— —— negative behaviour of proteid-cleavage sugar	40—41
Osazones	17
Osazone crystals from sugar of egg, photo-engraving of	207
—— —— —— —— of kidney, photo-engraving of	201
—— —— —— —— of lung, photo-engraving of	203
—— —— —— —— of muscle, photo-engravings of	194—196
—— —— —— —— of normal urine, photo-engravings of	183—185
—— —— —— —— of spleen, photo-engraving of	200
—— from cleavage sugar, after passage of primary potash cleavage product through a copper combination, photo-engraving of	42
—— —— —— —— from pepsin digestion of egg albumin, photo-engraving of	51
—— intermediate body, obtained in process of recovery of sugar from cleavage-sugar osazone, photo-engraving of	47
—— —— proteid-cleavage sugar, photo-engraving of	38
—— —— —— —— after recovery from lead compound, photo-engraving of	40
—— —— —— produced by direct action of sulphuric acid, photo-engravings of	45—46
—— —— sugar in beef tea, photo-engraving of	119
—— —— —— of blood of general circulation, photo-engraving of	158
—— —— —— of liver frozen instantly after death, photo-engraving of	138
—— —— —— of peptonised meat, photo-engraving of	120
—— —— —— of portal blood, photo-engraving of	107
—— —— —— recovered from cleavage sugar osazone, photo-engraving of	48
Papain, example of peptonising ferment in plant	23
Pancreas, sugar in	202
Pancreatic extract, experiments with	85—87
—— juice, amylolytic power of	84
Pasteur, observations on proteid formation in yeast	239
—— on growth of yeast cell by incorporation of sugar	53—54
—— researches upon yeast	3, 20, 53
Pepsin digestion of albumin, sugar produced in	50
Peptone, interference with reaction of copper test	43
Peptonised meat, presence of sugar in	119—120
Peptone, probable conjugation of, with carbohydrate in villi	242
Phenyl-hydrazine, application to product of action of sulphuric acid on albumin	43
—— behaviour of proteid-cleavage sugar with	39
—— —— of sugars with	17
—— reaction given by sugar of normal urine	182
—— test, value of	48—49
—— value of, as a sugar test, in presence of peptone	119
Photo-engraving of glucosazone from sugar of liver taken in ordinary way after death	140
—— of osazone crystals from sugar of egg	207
—— —— —— —— —— of lung	203
—— —— —— —— —— of kidney	201

	PAGE
Photo-engraving of osazone crystals from sugar of spleen	200
—— —— —— from cleavage-sugar after passage of primary potash cleavage product through a copper combination	42
—— —— —— —— from pepsin digestion of egg albumin	51
—— —— —— —— intermediate body obtained in process of recovery of sugar from cleavage-sugar osazone	47
—— —— —— —— sugar in beef-tea	119
—— —— —— —— —— of blood of general circulation	158
—— —— —— —— —— of liver frozen instantly after death	138
—— —— —— —— —— of peptonised meat	120
—— —— —— —— —— of portal blood	107
—— —— —— —— —— recovered from cleavage sugar osazone	48
—— of proteid-cleavage sugar osazone	38
—— —— —— —— —— after recovery from lead compound	40
—— of sugar titrating apparatus	76
Photo-engravings of osazone crystals from normal urine	183—185
—— —— —— muscle sugar	194—196
—— —— —— from proteid-cleavage sugar produced by direct action of sulphuric acid	45—46
Placenta and fœtus, sugar in	204
Portal blood, amount of sugar in	101—108
—— —— form of sugar in, after ingestion of cane-sugar	98—99, 108
—— —— in relation to ingested carbohydrates	101—108
—— —— low cupric oxide reducing power of sugar in	102
—— —— nature of sugar in	158
—— —— product of starch digestion in	89—90
—— —— sugar in, after fasting, after animal food, after starchy food, and after saccharine food	103—108
Potash, action on proteid	28—29
—— employment of, in process of glycogen (amylose carbohydrate) extraction and estimation	63
—— injection of, into liver for preventing *post-mortem* production of sugar	134
Proteid, action of potash on	28—29
—— -cleavage by sulphuric acid	43—49
—— —— carbohydrate, more resistant than glycogen to action of sulphuric acid	65
—— —— experiments, procedure adopted in	30
—— —— of, into fat, lactin, and casein of milk	259
—— —— origin of carbohydrate from	3—4
—— —— product by potash, copper compound of	35
—— —— —— —— diffusibility of	35
—— —— —— —— preparation and properties of	33—35
—— —— —— —— resemblance to animal gum	35, 41
—— —— —— —— strength of alcohol required for precipitation	33
—— —— sugar, behaviour with phenyl-hydrazine	39
—— —— —— combination with lead oxide	39
—— —— —— effects of different strengths of sulphuric acid in obtaining	37
—— —— —— in relation to fermentation and optical activity	40—41
—— —— —— osazone, melting point of	48

INDEX.

	PAGE
Proteid-cleavage sugar, preparation and properties of	36—38
—— —— —— produced by ferment action	491—5
Proteid, effect of water at elevated temperature on	32
—— formation from carbohydrate by cells of villi	121
—— glucoside constitution of, bearings in relation to diabetes	229
—— matter, glucoside constitution of	27—57
—— of haricot bean, cleavage experiments on	31
—— production from carbohydrate	233, 239—245
—— sugar contributing to formation of	3
—— synthetic formation by incorporation of carbohydrate	53—55
—— of blood serum, cleavage experiments on	31
Proteids, various, cleavage experiments on	31—32
Proteolytic action, not province of liver but of digestive ferments	121
Protoplasm, action on carbohydrate	233
—— effects conversion of carbohydrate into fat	246
—— functions of	251
—— mode of effecting transformation of carbohydrate into fat	258
—— participation in deposition of starch	55—56
—— vegetable, constructive power of	1, 2
Protoplasmic action, "metabolic" and "plastic" power of	24
—— and ferment actions opposed in their effects	260
—— —— —— —— reciprocal play of	24—26
Ptyalin, amylolytic ferment of saliva	82
—— example of unorganised ferment (amylolytic)	19

Qualitative and quantitative sugar testing	68—80
Quantitative determination of sugar by ammoniated cupric test	71—80
—— —— —— —— gravimetrically	69—70
—— —— —— —— of liver	136

Raffinose, constitution of	45
Recovery of cleavage sugar from its osazone	46—48
Ruminant, exceptional position of stomach and intestine in relation to cane-sugar inverting power	94, 96

Saccharoses, general description of	12—15
Saccharose, see cane-sugar.	
Sachs, on participation of protoplasm in deposition of starch	55—56
—— on peptonising ferments in the vegetable kingdom	23
—— on production of ferments in plants	25
Salicin, example of non-nitrogenous glucoside	27
Saliva, amylolytic ferment action of	82
Schwann, on "metabolic" and "plastic" power	24
Seeds, transmutations of carbohydrate in	25, 235
Seegen, on absence of difference between arterial and venous blood in relation to sugar	171
—— on amount of sugar in blood	162—163
Sodium carbonate, effect on action of amylolytic ferments	84—87

	PAGE
Sodium carbonate diminishes ferment action in liver substance	155—156
—— sulphate, use in preparing products for analysis	58—59, 62
—— —— not to be used in preparing blood for cane-sugar estimation	108
—— —— restraining influence on inverting action of sulphuric acid	67
Spleen, sugar in	200
Starch, action of alkalies, acids, and ferments on	9
—— action of saliva upon	82
—— chemistry and properties of	8—9
—— conversion into sugar necessary for absorption	81—82
—— deposition in plant	3—4
—— digestion of	81—90
—— digestion, product of, in portal blood	89—90
—— first visible primordial carbohydrate product	4
—— time required for conversion by sulphuric acid	66—67
Starchy food, portal blood after ingestion of	105—106
Stomach, absorption of sugar from	83
—— cane sugar inverting power of, large in ruminant	94, 96
—— —— —— —— slight in animals generally	94—96
Stomach-contents, cane sugar inverting power of	97—98
Succus entericus, alleged lactose-transforming power of	99
—— —— glucose-forming capacity of	87—89
Sucrose, see cane sugar.	
Sugar, amount and nature of, in portal blood after starch ingestion	89—90
—— —— —— in blood considered generally	158—165
—— —— —— in blood of general circulation	161
—— —— —— in general circulation not variable as in portal in relation to food	112
—— —— —— in normal urine	186—187
—— —— —— in pôrtal blood	101—108
—— aqueous extraction of, from blood	159
—— blood in relation to	157—177
—— Brücke's process for separation of, from normal urine	179—180
—— cane, changes of glucose and cane sugar in the growing plant	21—22
—— —— different glucoses in fresh and dead plant	41
—— change in nature and amount during incubation	208—210
—— contributing to formation of proteid	3
—— determination of nature of	59—60
—— derived from tissues in severe diabetes	229
—— from ingested carbohydrates, liver in relation to	109—131
—— from proteid of animal food not produced in liver but by digestive action	121
—— in arterial and in venous blood	166—171
—— in brain	204
—— in egg	206—210
—— in generative organs of fish and crustacea	205
—— in kidney	201
—— in liver at moment of death and afterwards, amount and nature of	136—139
—— in liver of cold-blooded animal	140—143
—— in lung	202—203
—— in muscle	194—199

INDEX. 279

	PAGE
Sugar in pancreas	202
—— in placenta and fœtus	204
—— in portal blood after fasting, after animal food, after starchy food, and after saccharine food	103—108
—— in portal blood, low cupric oxide reducing power of	102
—— in spleen	200
—— kind of, in urine after injection of glucose	236—238
—— liver in relation to	132—156
—— methods of, extracting for analysis	58—62
—— nature of, in blood	157—158
—— —— in portal blood	158
—— of blood increased under abnormal conditions	163
—— of liver, nature of	125
—— of normal urine, fermentability of	181
—— of tissues, process for ascertaining nature of	132
—— of urine stands in relation to that of blood	187—193
—— possible existence of unknown modifications or varieties of	44—45
—— *post-mortem* production in liver	133—140
—— presence of, in beef tea	118—119
—— —— in healthy urine	178—186
—— present in animal food	118—119
—— production of, in alcohol-coagulated liver-substance	147—154
—— —— in liver after removal and washing free from blood	146—147
—— —— in liver during life under certain abnormal conditions	145
—— production from the tissues in diabetes	263
—— quantitative determination of, by the ammoniated cupric test	71—80
—— question of disappearance from blood after withdrawal	171—176
—— rapid production of in liver after death	125—126, 144
—— same amount in liver as in other structures during life	144
—— standard amount in blood of general circulation	101
—— starch converted by digestion into	82
—— stoppage of, by the liver, evidence of, afforded by the blood	109—112
—— —— —— —— —— by the organ itself	112—118
—— testing, qualitative and quantitative	68—80
—— urine in relation to	178—193
Sugars, artificial, with 7, 8, and 9 carbon atoms	1
—— with less than 6 carbon atoms	1
Sulphuric acid, cleavage of proteid by	43—49
—— —— different strengths of, in relation to production of proteid-cleavage sugar	37
—— —— use of, in differentiation of carbohydrates	58—60, 62
Sun's rays, influence in production of organic compounds	1, 2
Synthesis of carbohydrates	5
—— —— in plant	2
Synthetic formation of proteid by incorporation of carbohydrates	53—55
Temperature, moderately elevated, effect upon disappearance of sugar from drawn blood	174—175
Testing for sugar, qualitative and uantitative	68—80

INDEX.

	PAGE
Tortoise, liver rich in glycogen	131
Transmutation of carbohydrates	233—239
—— in seeds and tubers	25—26
—— of sugar into glycogen a process of dehydration	117
Tuber, transmutation of carbohydrate in	26

Urine affords an index of the state, as regards sugar, of the blood 162
—— diabetic, glucose the form of sugar in, after ingestion of lactose 99
—— information derivable from, refutes glycogenic doctrine............ 225—227
—— in relation to sugar .... 178—193
—— kind of sugar present in, after injection of glucose................ 236—238
—— normal, amount of sugar in .... 186—187
—— —— behaviour with Fehling's solution under different circumstances.. 178
—— —— presence of sugar in... 178—186
—— sugar of, stands in relation to that of blood .... 187—193

Vaso-motor paralysis in connection with diabetes...................... 263
Vegetable kingdom the source of carbohydrates 1
Villi, cells of, as fat-formers from carbohydrate 248
—— —— as proteid-formers from carbohydrate .... 242—244
—— —— possess power of forming proteid and fat from carbohydrate 121
—— photo-engraving of appearance of cells of, after fasting............ 253
—— —— of appearance of cells of, after ingestion of oats............. 253
—— —— of micro-photograph representing appearance in section, after fasting 254, 256
—— —— —— —— after ingestion of oats............. 255, 257
Vitellin from yolk of egg. cleavage experiments on 31
Volumetric method of sugar determination 70

Water at elevated temperature, effect on proteid...................... 32

Yeast cell affords illustration of synthesis of proteid by incorporation of carbohydrate .... 53—54
—— affords illustration of dehydration by living matter .... 20—21
—— cells, illustrating conversion of carbohydrate into fat............ 246
—— —— illustrating proteid formation from carbohydrate 239
—— occurrence of glycogen in...................................... 10

Zoamylin, a more appropriate term than glycogen 9
Zymogen.. 19
Zymolysis, see Ferment action.

www.ingramcontent.com/pod-product-compliance
Lightning Source LLC
Chambersburg PA
CBHW032103230426
43672CB00009B/1622